Contents

A complete list of contents starts on page 212

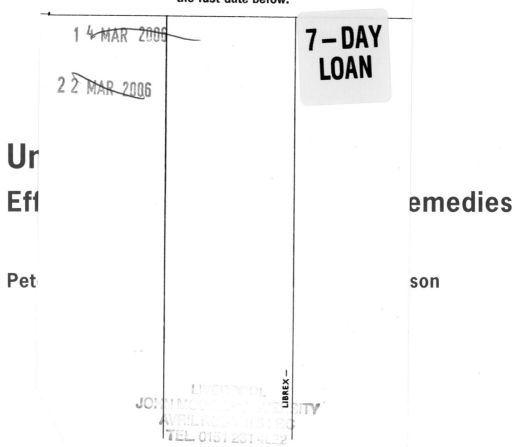

Books are to be returned on or before the last date below.

1 4 MAR 2006

2 2 MAR 2006

Un
Eff   emedies

Pet   son

BRE

ıg the future

BRE is the UK's leading centre of expertise on building and construction, and the prevention and control of fire. Contact BRE for information about its services, or for technical advice, at:
BRE, Garston, Watford WD25 9XX
Tel: 01923 664000
Fax: 01923 664098
email: enquiries@bre.co.uk
www.bre.co.uk

Details of BRE publications are available from:
www.brebookshop.com
or
IHS Rapidoc (BRE Bookshop)
Willoughby Road
Bracknell RG12 8DW
Tel: 01344 404407
Fax: 01344 714440
email:
brebookshop@ihsrapidoc.com

Published by BRE Bookshop
Requests to copy any part of this publication should be made to:
BRE Bookshop,
Building Research Establishment,
Watford WD25 9XX
Tel: 01923 664761
Fax: 01923 662477
email: brebookshop@emap.com

Chapter 1
Introduction

WHAT IS DAMPNESS?

In many situations the professional investigating the dampness will need an indication of the actual level of moisture within the structure. There is often a perceived need to know when a material is what might be called 'dry', although in absolute terms a porous material in a building will always retain some moisture, either from its own natural properties (for example hygroscopicity) or from the effects of water-absorbing (deliquescent) salts.

Even in a normal 'dry' building, there is always a surprising amount of water present in porous materials, most of which does no harm whatsoever. Although the amount varies widely, depending on the nature of the material and on the humidity of the surrounding air, the following figures indicate the range that may be expected in some common materials:
- plaster 0.2 – 1.0% wet weight;
- lightweight concrete >5%;
- timber 10 – 20%.

These amounts of moisture do little harm to the materials and such moisture is not usually regarded as dampness. That term is commonly reserved for conditions under which moisture is present in sufficient quantity either to become directly perceptible to sight or touch, or to cause deterioration in the decorations and eventually in the fabric of the building.

A building is considered to be damp only if the moisture becomes visible through discolouration and staining of finishes, or causes mould growth on surfaces, sulfate attack or frost damage, or even drips or puddles. All of these signs confirm that other damage may be occurring.

TYPES OF DAMPNESS

A high proportion of dampness problems turn out to be one of the big three:
- condensation;
- rain penetration;
- rising damp.

As these are so common, it is easy to overlook other causes but, before forming an opinion, these other causes of dampness should be investigated:
- construction moisture;
- pipe leakage;
- leakage at roofing features and abutments;
- spillage;
- ground and surface water;
- contaminating salts in solution.

Condensation

The causes of condensation are quite complex. When making an initial dampness diagnosis, there are several distinctive features to look for:

❐ Condensation normally occurs only in the coldest months of the year.
❐ Trouble starts on the coldest internal surfaces: external walls, particularly corners, single-glazed windows, cold-water pipes, wall-to-floor junctions, lintels and window reveals.
❐ Damp patches sometimes have definite edges on cold spots such as lintels; patches of damp or mould in exposed corners are crescent-shaped.
❐ Condensation occurs most often in rooms where large amounts of moisture are produced, such as kitchens and bathrooms, and in unheated rooms into which moisture has drifted.
❐ It is common in rooms where flueless paraffin or butane heaters or unvented tumble driers are in use, or clothes are frequently dried.
❐ It often concentrates in areas where air movement is restricted, such as behind furniture or inside cupboards on outside walls.

Gents

DUE TO CONDENSATION, THE FLOOR CAN BECOME VERY SLIPPY. PLEASE TAKE GREAT CARE

Under normal occupancy of a building in winter, temperature and vapour pressure falls from inside to outside. In multi-layered constructions, the gradients through the wall relating to both these properties depend on the relative thermal and vapour resistances of each layer. Interstitial condensation can result when the main thermal resistance of the wall is on the warm side of the the main vapour resistance. The extreme situation can arise when the outside of a wall has a dense render or an impermeable rain screen cladding, slowing down or preventing the escape of any vapour.

This may not matter; for example, condensation on the the outer leaf of a brick cavity brick wall will be negligible compared to normal wetting from rainfall. However, materials which can be susceptible to moisture, such as timber and timber-based products, or metals which can corrode, are much more vulnerable and can give rise to problems, especially structural members. The studs in a timber framed wall or metal reinforcement of a masonry wall may need special protection from dampness.

Rain penetration

Rain penetration occurs most often through walls exposed to the prevailing wet winds, usually south-westerly or southerly.

Even if rain penetration is certain to be the cause of the dampness, pinpointing the exact route that the rain is taking can be quite difficult. A damp patch on a ceiling could be due to a missing roof tile or to a faulty flashing some distance from the patch. Materials in parapets and chimneys can collect rainwater and deliver it to other parts of the building below roof level, unless they have adequate DPCs and flashings. Blocked or defective rainwater goods can lead to damp patches on walls that appear to be straightforward rain penetration.

Rising damp

The results of rising damp in walls leave characteristic signs. There is usually a fairly regular, horizontal tide mark, up to a couple of metres above the floor. Below it, the wall is discoloured with general darkening and patchiness; there may be mould growth and loose wallpaper. Hygroscopic salts brought up from the ground tend to concentrate in the tide-mark. In severe cases, it may cause rot in skirtings or dados.

If there is a physical DPC, it is unlikely to have failed. But it could be bridged by pointing or rendering, or by soil, paving or rubbish heaped against the wall outside, or by plaster inside. In a cavity wall it could be bridged by a build-up of mortar droppings.

If there is no DPC, the presence of rising damp must be confirmed by correct diagnosis before deciding to put in a remedial DPC.

Rising damp can affect solid and suspended floors. In solid floors, it is occasionally due to a faulty or missing DPM. Dampness in suspended timber floors often becomes evident by the discovery of wet or dry rot, starting in the joist ends.

Salts show extent of rising damp in an agricultural building

Construction moisture

In a wholly or partly new building, the fabric contains water used in concrete, mortar and plaster. In a typical brick-and-block, semi-detached house, about 8000 litres of water can be used for mixing. This can take a long time to dry out; for example, a 150 mm-thick floor slab may take about a year. In addition, bad weather during construction may have saturated the building before it was closed in. Water can also be trapped in the fabric of older buildings which have been open to the weather during repair.

Leaking pipes

Over time, even a small leak in a water supply, central heating, drainage pipe or rainwater goods can cause extensive dampness, often some distance from the leak. The dampness can easily be mistaken for rising damp, rain penetration or condensation.

Leaks at roofing features and abutments

Blocked valley gutters and downpipes can cause rainwater to pond and overspill the flashings. Parapets and chimneys can become extremely wet and, in the absence of effective damp-proofing, water will drain downwards to other parts of the building, showing as damp patches in rooms below.

Spills

Persistent or recurring spillages can occur from tanks, cisterns, washing machines and dishwashers. Frequent floor washing can also cause problems, for example in kitchens in institutional buildings. Water running through cracks or joints in an impervious floor covering can spread underneath and may reach areas where drying out is either impossible or which may take a considerable time to complete.

Ground and surface water

Water can seep into ground floors or basements from ground or surface water, or from repeated flooding.

Contaminating salts

Walls and floors can become contaminated by hygroscopic salts causing damp patches to form.

Where is dampness apparent?

In some situations, dampness is a surface effect, such as condensation on the underside of pitched metal roof sheets or on single-glazed windows; in other places, the mass of the material can become saturated, such as a floor slab and the base of walls soaked over a long period by a leaking central heating pipe buried in the screed.

Some effects of dampness are easily visible: damaged decorations resulting from long-established rising damp; swollen timber and loss of adhesion of a paint film from a thorough wetting; outbreaks of black mould growth coupled with rain penetration; or a similar mould growth pattern from thermal bridging. Other effects may take time to show: distortion of floor sheeting material by moisture beneath (osmosis effects) and dampness beneath an insulated chipboard floor found only when the chipboard disintegrates.

For many years, BRE has been involved in studying these effects, both when investigators are carrying out short-term site inspections and examining real situations in existing buildings, or are studying phenomena over a long period, such as the drying of mass concrete structures.

There is no universal treatment for curing dampness which has become a nuisance and rarely is it easy. Each problem of dampness must be considered individually and the cause correctly diagnosed before a method to cure the defect can be prescribed. Even when a cause that would appear to account for the dampness has been found, it is wise to continue the examination until it is reasonably certain that there are no other contributory causes. If any of these are missed, the cure for the main cause may accentuate the dampness from these other causes and the treatment will then be regarded as a failure.

RECORDS OF DAMPNESS RELATED PROBLEMS

BRE Advisory Service records

Dampness in buildings has been a continuing feature over the years. In one form or another it was found to be the main problem in around half of the cases involved in each of three separate analyses, each of over 500 site investigations carried out by staff of the BRE Advisory Service during the sample periods 1970–74, 1979–82 and 1987–89.

Rain penetration

Rain penetration featured in 25% of the 510 occurrences during the period 1970–74; 27% of the 518 occurrences during the period during 1979–82 and 22% of the 520 occurrences during the period 1987–89, an average of about one in four of all investigations.

During the period 1970–1974, of the 510 investigations carried out, about one in five concerned the external wall. Rain penetration was the the defect most frequently investigated. About half the number of cases occurred in cavity filled walls via DPMs and trays, and the other half in solid walls, concrete cladding and other kinds of external wall. Rain penetration actually through windows, as opposed to the window-to-wall joints, also occurred in a significant number of cases.

Condensation

Condensation featured in 17% of the 510 occurrences during the period 1970–74; 15% of the 518 occurrences during the period 1979–82 and 17% of the 520 occurrences during the period 1987–89, an average of about one in six of all investigations.

In 1988, BRE investigators carried out postal and interview surveys in some one-bedroom and bedsitting-room homes. The main aim was to examine problems relating to condensation. Half the homes studied had enough condensation to cause pools of water on the window sills, and one in six had sufficient mould growth to cause damage to plaster or woodwork. Problems reported by occupants were strongly correlated with observations made by interviewers. In these small homes, condensation problems were related to location in the UK (being worse in warmer areas), age of respondent (retired people having fewest problems), household size, insulation standards, home heating (particularly the use of bottled gas) and air movement within the home, but not ventilation habits. Condensation and mould growth certainly used to be widespread problems in all housing sectors, but especially so in tenanted accommodation. In many cases it was difficult to identify the underlying cause; this was often complicated by social issues.

Entrapped water

Entrapped water featured in 5% of the 510 occurrences during the period 1970–74; 3% of the 518 occurrences during the period 1979–82 and 6% of the 520 occurrences during the period 1987–89, an average of about one in 20 of all investigations.

Rising damp

Rising damp featured in 5% of the 510 occurrences during the period 1970–74; 4% of the 518 occurrences during the period 1979–82 and 5% of the 520 occurrences during the period 1987–89, an average of about one in 20 of all investigations.

BRE Defects database records

BRE inspections

In separate studies of the quality achieved in new-build house construction, carried out during the early 1980s, and replicated during the late 1980s, each covering upwards of 1000 dwellings, risk of dampness occurring in one form or another was identified on many of the sites. The resulting database recorded actual inspections by BRE investigators or by consultants working under BRE supervision. The research recorded non-compliance with requirements whatever their origin, Building Regulations, codes of practice, British and industry standards or other authoritative requirements, which could lead to dampness.

In the earlier study, potential weathertightness problems featured in 16% of cases, rising damp in 4%, and condensation in 5%. Entrapped water could not be measured. About one in four of all cases related to dampness.

In the later study, potential weathertightness problems featured in 12% of cases, rising damp in 4%, and condensation in 5%. Entrapped water could not be measured although the investigators were surprised to find extensive mould growth in one brand-new house which had yet to be occupied; it was suspected to have been caused by entrapped construction water. The study showed roughly similar results to the earlier study; this suggested that there had been no change over the intervening eight years. The UK construction industry has always been slow to learn the lessons of its past mistakes – see BRE IP 3/93.

Public sector experiences

In parallel with these studies, public sector building owners were questioned in 1983 and again in 1989 about their experiences of problems with their housing stocks; from the results it was possible to place the problems in order of importance. 115 authorities responded to the invitation.

Tables 1.1 and 1.2 show that dampness related problems were the top five in public sector new-build housing in 1989, and the top seven in rehabilitated housing. There were hardly any changes from the 1983 figures, although there were some differences in the rank order. The more recent experience of the BRE Advisory Service indicates that current problems do not differ significantly from these figures.

House condition surveys

The main source of information on the condition of the UK housing stock is the four house condition surveys carried out every five years in England, Wales, Scotland and Northern Ireland. These are a detailed survey of about 45,000 dwellings in a representative sample, with a follow-up interview with the householders by a market research firm, to establish the ways they use their house, their attitudes to such issues as energy conservation and any problems they experience. The most recently published information is from the 1996 English and Scottish Surveys.

England

The 1996 *English house condition survey* shows that 3.9% of dwellings were affected by rising damp and 6.0% by penetrating dampness. These are both much more common in older buildings which are less likely to have a complete DPC and more likely to have solid walls – Figure 1.1.

Table 1.1 Public sector housing survey: new-build Rankings in 1989 and 1983 compared				
1989	**Performance of building element**	**Number of mentions**	**% of 115**	**1983**
1	Weathertightness of windows and doors	33	29	3
2	Surface condensation on brick external walls	29	25	1
3	Weathertightness of flat roofs	25	22	5/6
4	Rising damp in brick external walls	22	19	2
5	Weathertightness of pitched roofs	21	18	5/6
6	Surface condensation on windows and doors	19	17	9/10
7	Surface condensation on other external walls	17	15	4
8	Strength and stability of foundations/basements	14	12	9/10
9/10	Durability of windows and doors	12	10	9/10
9/10	Durability of fixtures and fittings	12	10	-
-	Weathertightness of brick/block external walls	-	-	7
	Rising damp in other external walls	-	-	8

Table 1.2 Public sector housing survey: rehabilitated housing Rankings in 1989 and 1983 compared				
1989	**Performance of building element**	**Number of mentions**	**% of 115**	**1983**
1	Weathertightness of flat roofs	59	51	4
2	Weathertightness of windows and doors	54	47	1
3	Rising damp in brick block external walls	54	47	3
4	Surface condensation on brick external walls	51	44	2
5	Weathertightness of pitched roofs	46	40	5
6	Strength and stability of foundations/basements	37	32	6/7
7	Strength and stability of external brick block walls	33	29	10
8	Weathertightness of brick block external walls	32	28	8/9
9/10	Surface condensation on other external walls	32	28	6/7
9/10	Surface condensation on windows and doors	31	27	-
-	Rising damp in brick block separating walls	-	-	8/9

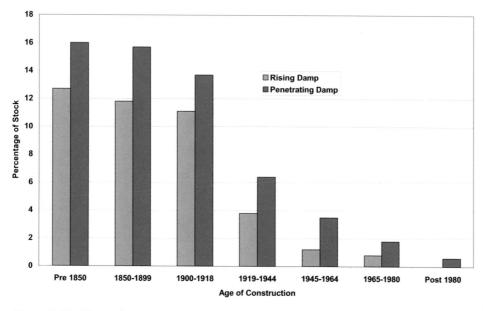

Figure 1.1 Incidence of rising and penetrating dampness by age – *from The 1996 English house condition surve*

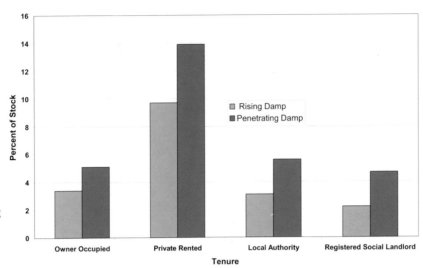

Figure 1.2 Incidence of rising and penetrating dampness by tenure – *from The 1996 English house condition survey*

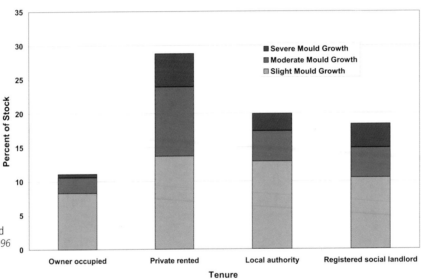

Figure 1.3 Incidence of mould growth by tenure – *from The 1996 English house condition survey*

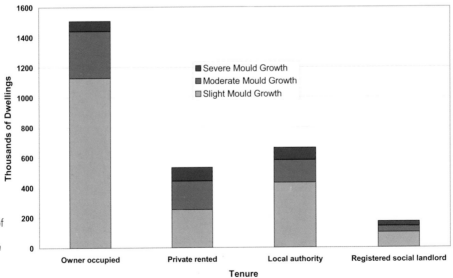

Figure 1.4 Numbers of dwellings with mould growth by tenure – *from The 1996 English house condition survey*

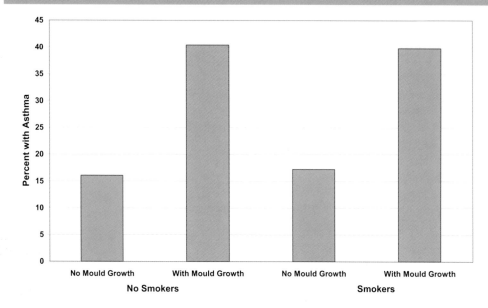

Figure 1.5 Percentage of households with at least one person complaining of asthma, by mould growth and presence of smokers – *from The 1996 English house condition survey*

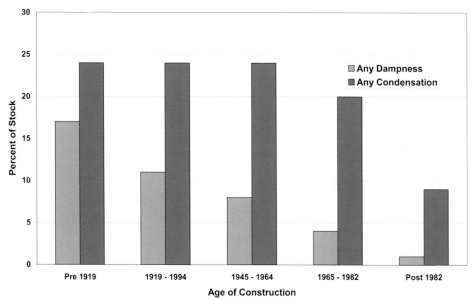

Figure 1.6 Dampness in the older housing stock in Scotland – *from The 1996 Scottish house condition survey*

Problems are also concentrated in private rented housing, which is often older and in poorer repair that the rest of the stock – Figure 1.2.

Altogether, 14.6% of dwellings in the EHCS had some degree of mould growth on the walls of furnishings. Rented housing is more severely affected with the most severely affected part of the housing stock being the private rented sector – see Figure 1.3 – which is older, in poorer repair and occupied by more low income households than the other parts.

Because so many dwellings in England are now owned by their occupants, 52% of mould problems occur in owner-occupied housing – Figure 1.4.

Figure 1.5 shows particularly disturbing evidence of the effect of mould on the health of occupants.

The percentage of households with at least one person suffering from asthma more than doubles in houses with mould growth, even when allowance is made for smoking, the other major cause of respiratory problems.

Scotland

The picture from the 1996 Scottish house condition survey is broadly similar; about 25% of dwellings suffered from dampness or condensation. Dampness and condensation affected 4% of dwellings, dampness alone a further 4% and condensation alone 17%. As in England, dampness was concentrated in the older housing stock – see Figure 1.6. The incidence of condensation was more or less constant up to the 1980s, when improved insulation standards should be expected to reduce the risk.

Comparable figures for Wales and Northern Ireland were not available at time of writing.

CHANGES IN LIFESTYLE AND CONSTRUCTION

Changes in domestic lifestyles

Enormous changes have taken place in domestic lifestyles in the past half-century, especially in heating, the nature and costs of fuels, and a reduction in ventilation rates. In houses heated by open coal fires and those where the building byelaws called for the installation of air bricks, the risk of condensation was low. Following the increases in the costs of fuels, occupiers naturally tried to save money by blocking up ventilation. If the household were out during the day, they wanted instant heat on their return, sometimes using paraffin or LPG-fuelled heaters, with appaling levels of condensation on the cold structure of the dwelling.

There have also been considerable changes with domestic clothes' washing arrangements. Solid-fuelled, open-top wash boilers were still being installed in houses in the early 1950s, and gas or electric boilers were used elsewhere, producing large quantities of water vapour, even steam, within the dwelling. Now, of course, the automatic washing machine is common, and some occupiers have learnt to ventilate tumble driers to the external air.

Changes in external walling practice

Brick walls were almost invariably built in a single leaf until the gradual introduction of cavity walls from early Victorian times. In single-leaf domestic construction, rendering was common until the 1939–45 war, and much of the ribbon housing development of the late inter-war period was still being built in 9-inch brickwork. In areas subject to heavy driving rain it was not unusual to use tile or slate hanging to ensure the wall was watertight.

Damp-proof courses

Damp-proof courses were not common in domestic construction until required by the Public Health Act 1875, and the walls were afforded further protection against dampness by, for example, overhangs and drip moulds which directed rainwater away from the face of the building.

Cavity walls

Cavity walls, at first in twin leaves of brick with facings in the outside leaf and commons on the inside, became standard practice for domestic work only after the 1939–45 war, though they had occasionally been used in certain parts of the country 100 years earlier. Blocks, at first made with coke breeze or clinker aggregates, then gradually replaced the commons in the internal leaves. The main aim of the cavity wall was to improve weathertightness which, provided the workmanship was satisfactory, it did, and the incidence of rain penetration could be reduced. Although it was realised that the thermal insulation value of the wall had been improved, such improvements were marginal when considered in today's terms.

Doors and windows

Doors and windows have improved significantly in performance from the double-hung sashes set into half-brick reveals, both as regards thermal insulation and consequent reduction of condensation, and also in weathertightness and reduction of rain penetration. There was, however, an hiatus in the post-World War II developments, when steel and timber casements became popular, set almost flush with the external face; these were more vulnerable to rain penetration. This seemed to occur more in England and Wales than in Scotland, where the tradition of recessing the windows for better protection was retained. The comparatively recent developments of plastics and other high performance units have, however, given significant improvements both in thermal insulation, reduction in condensation, and in improved weathertightness.

Non-domestic construction techniques

In the second half of the twentieth century, many new external wall techniques were introduced; for example, curtain walling and rain-screen cladding used as new construction or simply to over-clad leaking traditionally-built facades. For the most part, these techniques led to relatively impermeable outer skins of metal, man-made sheet materials, and large areas of glass. Any gaps or holes in this skin could lead to water being pumped through by wind pressures and suctions; any missing thermal insulation could lead to condensation risk. Any misunderstanding on the part of either designer or builder of the required characteristics of these new techniques could lead to potential problems, and the files of the BRE Advisory Service are full of such examples.

Changes in floor construction practice

Many buildings built before the first decade of the 20th century were constructed with solid floors of puddled clay or of quarries laid on ash beds, laid without benefit of DPC. In theory at least, these were vulnerable to rising damp. In practice, however, rising damp was rarely seen.

Suspended timber floors

The risk of rising damp through the floor should have been entirely removed with the introduction of suspended timber floors which became popular in late Victorian times. Some, however, were laid without DPCs on the sleeper walls, leading to rising damp and to consequent rot in the joists. Many domestic buildings built during the inter-war period had suspended timber floors but the shortage of timber in the early post-World War II period led to a change in common practice and a reversion to solid floors.

Solid floors

This change from suspended back again to solid floors on a major scale following the 1939–45 war took place in England, though not to anything like the same extent in Scotland, Wales and Northern Ireland. At this time a variety of floor finishes was used, the most common being magnesite, pitchmastic, mastic asphalt and linoleum laid directly on concrete bases. But with the introduction of thermoplastic tiles around 1946, the use of all except linoleum had disappeared from domestic construction by around 1960. Because thermoplastic tiles and the solvent bitumen adhesives used to fix them were moderately tolerant of moisture rising from below, it was common to lay this type of flooring directly onto a concrete base, usually four inches (100 mm) thick without any DPM or screed. Except on very wet sites, this construction worked well. The provision of DPMs below concrete slabs did not become common until sheet polyethylene became generally available in suitable size and thickness around 1960. Prior to that date, where DPMs were required to protect moisture-sensitive floorings, they were formed by applying various brush or hot-applied tar and bitumen-based products to the top of the base concrete and covering it with screed or board based products. The use of pitchmastic and mastic asphalt as both screed and surface DPM was also common.

It is only within the last few years that precautions have been taken to increase the thermal insulation at the perimeters of solid floors in new construction, with the consequent reduction of condensation risk on the lower temperature of the concrete near to the external walls.

Changes in roof construction practice

Pitched roofs

Various structures and coverings have formed most roofs in the UK since medieval times and a fair majority of the nation's stock of buildings still have pitched roofs of one kind or another. Maintenance of the weathertightness of the coverings of such roofs is for the most part straightforward. However, in the case of the underlying structure, for example of heritage buildings, there can be problems which need specialist advice.

Sheeted, low-pitched or almost flat roofs have become popular within the last twenty years. These roofs often have steeply pitched tiled perimeters; and they are not entirely without their problems.

In the domestic short-span field there has been an almost complete swing away from the strutted purlin roofs of inter-war and early post-war years towards trussed rafter roofs. These came into widespread use in 1964. Since the mid-1980s, a further noticeable swing in fashion has been from hips to plain gables and back to hips. Provided the construction is satisfactory, there should be virtually no difference in performance so far as dampness is concerned, even considering the potential corrosion of steel trussed rafter plates. What does make a difference, though, is the use of clipped eaves, which offer no protection from rainfall to the wall below.

Another tendency which has become apparent is that of increasing complexity of the geometry of buildings, and more especially of roofs. It is also clear that defects increase in direct proportion to increases in complexity of geometry of the surfaces of buildings; in other words at the intersections of different planes. These intersections provide many situations where the roof continuity is interrupted, and which therefore become vulnerable to rain penetration.

Flat roofs

In the 1920s there was a swing in architectural fashion away from the pitched roof in favour of the flat roof. The flat roof was also the only conceivable solution for some of the convoluted plan forms that were adopted. It is the poor performance of some of these, particularly system-built roofs, coupled with lack of tolerance to thermal and moisture movement of structures leading to cracking of the membranes, that gave flat roofing a bad name.

The real difficulty came with built-up bitumen felt flat roofs which failed in large numbers in the 1950s and 1960s. Particularly when exacerbated by poor maintenance strategies, owners of large building stocks with these roofs, such as county councils and the Property Services Agency, found themselves responsible for substantial costs for their repair. The roofs which failed were largely those using organic felts fully bonded to the deck immediately below them. Consequent thermal movement of the substrate from exposure to solar radiation caused the waterproof membrane to split and break down after a relatively short life.

With the introduction of newer materials and better practice, there should be a progressive reduction of the incidence of defects in newly built flat roofs.

This surface reading may not be representative of the overall moisture content of the component. Naturally occurring mineral salts, timber preservatives and embedded metals can affect the electrical resistance and, therefore, the measured moisture content. Preliminary investigation or laboratory tests may be needed to quantify this level of interference. There are two principal components in interpreting moisture data from timber components:

❏ the actual level of moisture (% volume reading);
❏ the change in this moisture level over time.

There will be seasonal variation in moisture content. Consider, for example, timber rafters and purlins within a conventional ventilated roof-space and the timber studs within the external walls of a timber framed house. The moisture content rises during the wetter, winter months. But moisture contents reduce in spring and summer. In some situations, moisture contents above 20% may be tolerated provided they are not sustained for more than a few weeks. These figures should be used simply for comparison rather than as key reference data – see Good Repair Guide 33.

Where moisture content measurements are repeated over a period of time (such as long-term monitoring of historic buildings), it is the changes in moisture content, not the absolute level, which are significant. For example, moisture contents rising, rather than falling, during spring and summer suggests that there is some form of moisture storage within the building fabric. Moisture stored within the masonry may evaporate during the warmer months and find its way into the colder roof space where it condenses.

Higher moisture content levels towards the outer face of the wall suggest wetting from wind-driven rain. As the outer face is wetted, and remains wet for some time, moisture is absorbed further into the the wall creating a reducing moisture content through its depth. If staining on internal surfaces is suspected to be caused by rainwater penetrating the thickness of the wall, the moisture content profile across the full depth of the wall is likely to be relatively flat, but at a high moisture content level.

The extent of exposure to wind-driven rain can be determined by comparing the moisture content profiles up the height and across the face of a building. Take into account the protection afforded by such details as window sills, copes, and overhangs. In general, the moisture content of the outer parts of the fabric increases with height up the building and is generally higher towards the outer edges of each facade. A reverse in this

Measuring moisture content of a trussed rafter roof

By carefully planning a measuring pattern across the roof area and repeating the measurements every few weeks, it should be possible to determine:

❏ the extent of any moisture within the roof timbers;
❏ if the moisture pattern is localised or influenced by orientation;
❏ whether the roof timbers are wetting or drying;
❏ whether the moisture is penetrating rain or condensing vapour.

Select three measuring locations: a truss at each end of the roof (say three trusses in) and a truss in the centre of the roof. Take moisture contents 250 – 300 mm from the end of each structural element (rafter, joist and web). At each location, take readings about 25 mm from the outer edge of the element. In the rafter, take measurements 25 mm from any sarking material. If the sarking is timber boarding, also take measurements in the boarding. This gives a moisture profile across the whole roof.

pattern may indicate localised moisture, such as rainwater splashing back onto the wall at low level, damaged rainwater pipe or rising ground moisture.

Take care when interpreting exceptionally high moisture contents towards the outside of the wall. A recent period of high rainfall could produce a moisture level three or four times that of the remainder of the wall but this would be a transient effect. When it becomes warmer and drier, the wall will dry naturally over a few days. Meteorological data for a period prior to the sample collection is an important part of the investigation.

Increased moisture levels towards the inside of a building generally indicate extreme cases of surface condensation but ensure that other sources of moisture are not influencing the measured results. Leaking water services, hygroscopic salts, wetting following flooding or rising ground moisture can provide similar results.

Higher moisture contents towards the middle of the wall are typical of most masonry walls during the spring/summer drying phase. During the winter, the overall moisture content of the wall and, in particular, the outer portions, would be expected to rise owing to wind-driven rain and condensation. From March to September, the drier months, the moisture content of the wall reduces as moisture evaporates from both the inner and outer faces.

Resistance gauges

These measure electrical resistance between electrodes embedded in a porous sample material which is in turn embedded in the test material. They do not directly measure the resistance of the test material itself but do so indirectly by absorption of moisture into the gauge material. They are more accurately described as absorption/resistance gauges.

There are problems with calibration. Although resistance varies uniquely with moisture content and temperature, the gauges are not equally sensitive at all moisture contents. Furthermore, distribution of water between the ancillary equipment and the test material depends on pore size distribution of the test material and on its moisture content.

Alternating current measuring instruments are sometimes used and several have been marketed for this purpose but they are essentially a research tool, rather than devices that can be used by a surveyor. Their life has been questioned but some inserted into a limestone wall at Gloucester Cathedral were found in excellent condition after not being used for 12 years.

Microwave techniques

Certain methods of moisture measurement depend on changes in the dielectric properties of porous materials with changes in the moisture content. Capacitance methods use the technique but microwave techniques use the propagation of electromagnetic radiation.

Equipment consists of a signal generator and receiver placed either side of a wall, although reflection methods allow tests only from one side. Commercial equipment is available, although performance claims are often optimistic. Good correlation between moisture content and attenuation can be obtained with homogeneous materials and, with careful calibration, results can be used with confidence. However, with non-homogeneous materials, such as the more usual sample of brick/mortar walls, the results are disappointing.

Capacitance methods

Dielectric properties of porous materials can be measured using conventional electrical circuitry and devices based on two electrode capacitors. Meters have been available for a long time and have been used on loose materials and for conveyer belt monitoring.

Commercial instruments for the building industry use a small, flat-plate measuring head with two conducting rings connected to a separate unit containing the circuitry. Examples include the *Aucon* and the *Sovereign* meters. Examined in tests similar to those used for resistance meters, reasonable results were obtained for pure water and low concentrations of salts with a flattening of the curve at high moisture content readings indicating insensitivity as saturation approaches. Meter readings are of little value with significant salt contents.

A further problem is with their use on rough surfaces. As well as giving spurious readings, there is a risk of damage to the measuring surface. BRE has made much use of this type of meter looking for dampness behind apparently dry wallpaper, when it was suspected that water has crossed a cavity fill material to wet the inner leaf.

Physical sampling by independent cores

This is relatively specialised. Using a diamond-core cutter, a 25 mm core is drilled from the wall; a rubber collar is then placed over both ends of the core so that it can be slid back into its original position. Before it is inserted, a wire is passed around the back of the core to facilitate subsequent removal.

The sample is usually taken to the laboratory, where it is sliced up, weighed and dried to provide a moisture content profile through the wall. The sample is then re-assembled and reinserted into the wall and left for a period of time. At intervals, the sample can be removed, re-weighed and the moisture content determined.

This is a quick and accurate way of determining moisture content but can be used only where moisture movement is perpendicular to the faces of the structure, such as rain penetration or drying out. It can be carried out on most masonry materials, is suited to the more technical or academic applications and is useful for long-term monitoring of historic buildings. The cores must be assembled carefully to ensure that there are no air gaps between the core slices and as small as possible between the core and the surrounding material. Any air gaps would prevent the core sample behaving as a component part of the structure. BRE experiments demonstrate that equilibrium is established fairly readily between the core and the wall even though the core and the wall are not in close all-round contact.

Drilled samples

The basis of this method is to drill out damp masonry or mortar and measure moisture content and hygroscopic moisture content. Walls may contain considerable quantities of hygroscopic salts so hygroscopicity should be measured to see whether the wall could have absorbed from the atmosphere the quantity of water found in the samples. On-site moisture content can be established by using a carbide meter, a commercial piece of equipment where the damp drillings and carbide are mixed in a pressure vessel. A gauge measures the pressure generated and the calibration is directly related to moisture content. If several samples are taken, it is more convenient to test in the laboratory and establish both moisture content and hygroscopic moisture content.

The advantages of the method are:
- ❐ it is independent of salts;
- ❐ it measures moisture content within the material rather than only in the surface layer;
- ❐ a moisture profile can be established by drilling in stages;
- ❐ measurements are made from only one side of the wall;
- ❐ it can be used on a wall where previous preparations, such as building in probes, is not possible;
- ❐ the equipment is inexpensive and may be already available, for example a chemical balance and desiccator (or a carbide meter can be purchased).

Samples can be gathered through the full depth of a masonry wall. In particularly thick walls, drillings at 25 mm or 50 mm increments can create a moisture content profile through the depth of the wall. Drilling is stopped 5 – 10 mm short of the inner face of each leaf to prevent penetrating the cavity and losing the dust sample. This staged drilling technique can be achieved by marking along the length of the drill bit or by using a depth gauge attachment.

The main disadvantage is that the method is semi-destructive, requiring at least a series of 9 mm holes or, if the wall is plastered, a chase cut out to locate the mortar joints. In practice, building owners are usually content for samples to be taken so that a definitive diagnosis can be made.

Figure 3.1 Equipment used in drilling

Figure 3.2 The complete calcium carbide kit including scales for weighing samples and pressure vessel (*right hand side*)

Drilling method

The technique is generally used on masonry materials such as brick, stone and concrete blockwork walling. It requires some basic equipment on site: a standard percussion drill, a 9 – 12 mm masonry drill bit, a device to catch the drill dust (some form of chute), a scraper and small bottles with stoppers in which to collect the dust sample – Figure 3.1.

It is better to drill into the mortar because the bricks will have a lower moisture content. Collect the sample in a stoppered bottle for subsequent laboratory tests. About two grams is sufficient, or six grams if the carbide meter is to be used. Take successive samples up the height of the wall and plot a graph of moisture content against height.

Measuring by calcium carbide technique

This is a quick and relatively accurate way to determine moisture content on site. By inserting individual samples with calcium carbide into a pressurised vessel, an almost instant reading of moisture content can be produced – Figure 3.2. Moisture in the drilled sample reacts with the calcium carbide to produce acetylene gas. The subsequent gas pressure indicates moisture content.

Measuring by weighing

For most situations, hygroscopicity can be measured at 75% relative humidity; for most walls the effective relative humidity is less than 75% and it is easy to provide a relative humidity of 75% by using a saturated solution of common salt (NaCl). Place a closable vessel containing the sample over, but not in contact with, the solution of common salt. A desiccator or an enclosed space similar to that shown in Figure 3.3 is suitable.

Figure 3.3 Weighing sample

Using a balance:

1 Collect the samples in stoppered bottles.
2 Shake the bottle well before removing stopper.
   Spread about 2 grams of the sample from the bottle onto a previously weighed Petri dish or watch glass about 40 mm diameter (weight W_o) and weigh immediately (weight W_w).
3 Place immediately in enclosure at 75 per cent relative humidity.
   Leave for a time: overnight is sufficient if the layer is only 1 – 2 mm deep.
4 Reweigh (W_{75}).
5 Place in an oven at about 100°C for about one hour. Remove and allow to cool.
   Reweigh soon after cooling (weight W_d). Plaster samples must not be dried out above 35°C.
6 Calculate:

Hygroscopic moisture content at 75 per cent RH (HMC) =

$$100 \ \frac{W_{75} - W_d}{W_{75} - W_o} \ \% \text{ wet wt}$$

Moisture content of sample when found (MC) =

$$100 \ \frac{W_w - W_d}{W_w - W_o} \ \% \text{ wet wt}$$

This procedure cuts down weighing to a minimum; if the moisture content is required more quickly, leave part of the sample in the bottle or put it on an additional dish and find the moisture content by weighing and oven drying.

The formal procedure is specified in BS EN ISO 12572.

Accuracy of the methods

A series of tests compared the drilling method using oven drying techniques and the carbide meter. The tests also demonstrated that the heat produced by a sharp drill bit does not cause evaporation of water from the sample, provided the material is not too hard.

This method satisfies more of the requirements of the 'ideal method' than any of the other techniques; it is sufficiently accurate for experimental work and can equally well be used for site investigations.

Calculating results by laboratory weighing

The method outlined above gives moisture contents expressed as % wet weight and is the convention commonly used in the damp-proofing industry.

The moisture content can also be expressed as % dry weight using:

$$\frac{(\text{Wet weight - dry weight}) \times 100}{\text{dry weight}}$$

Subtract the weight of any sample bottle from the wet and dry weights.

The calculated figure represents the moisture content by weight of the dry material. The moisture content of materials is often expressed 'by weight'. However, when comparing the moisture content of different materials it is generally more appropriate to express the moisture content 'by volume'. The moisture content by volume is calculated using:

Moisture content by weight x specific gravity of material

Calculate specific gravity by dividing the density (kg/m^3) by 1000.

Table 3.1 gives typical specific gravities for common building materials.

Table 3.1 Typical specific gravities for common building materials	
Material	**Specific gravity**
Lightweight concrete	0.6
Medium density concrete	1.1
Dense concrete	2.2
Common brick	1.8
Engineering brick	2.0
Plasterboard	0.9
Plaster	0.8
Render	1.5
Sandstone	2.0
Slate	2.7
Limestone	2.2
Reconstituted stone	1.7
Concrete roofing tile	2.1
Clay roofing tile	1.9

Moisture contents (%) at which action may be required

When dampness is suspected, accurate measurement of the moisture content of materials which may be subject to deterioration is essential. With these measurements, you may be able to reach a decision on whether remedial action is necessary and, if so, form some idea of the urgency with which the work should be undertaken. Table 3.2 gives a selection of values for typical materials, although their situation within a building will have some bearing on whether action will be necessary.

There was a net gain in moisture content throughout each set of measurements. Gains were small and consistent with there being three persons in the room at all times, and sometimes five. About 7.2 kg of moisture are generated each day by domestic activities so it must be realised that, unless means are taken to get rid of this moisture, high humidities and condensation are inevitable.

The level of heating seemed reasonable but the electricity consumption was well below that predicted by the local supplier for this type of property. With lower temperatures, the fabric of the building will become colder, so when the heating is switched on, condensation is more likely.

The major part of the lounge extended beyond the main part of the building with external walls on two sides, the third being a cavity party wall. The heat loss from this lounge could be expected to be quite high. Both the kitchen and bathroom were on the west side and the action of the prevailing wind would tend to move moisture across the building rather than out through the windows. An extract fan was later fitted in the kitchen; this helped to remove the moisture at source.

Air movement in the lounge was tested using smoke: there was little with the windows and doors closed, smoke released at ceiling level stayed close to the ceiling, just rolling backwards and forwards with no particular pattern. The windows were tight fitting and the patio door was sealed with adhesive tape. Even with reasonable natural ventilation, the close proximity of the door and windows meant that there would be little inducement for air in the far corner of the room to move. The patio was surrounded by a high wall, which further inhibited natural ventilation. A smoke test carried out with the top light windows open showed that the influence of these windows in ventilation was limited.

Conditions measured in January gave a dry bulb temperature of 9.3°C, RH of 84%, indicating a dewpoint of 6.5°C. So when the temperature of the roof drops in cold weather, condensation is bound to occur.

Conclusions

The first problem is the roof; the movement of moisture vapour from a high vapour pressure zone below the ceiling into the left space, coupled with inadequate loft ventilation, results in high levels of humidity. The roof temperature drops during the night, and condensation forms on the underside of the felt.

The second problem of condensation on the tops of all the walls is again caused by air dewpoint temperature being above the wall surface temperature. As the wall temperatures are not at an unreasonable level, the way in which the ceiling heating establishes a layer of high humidity air just below the ceiling must be the cause of the problem. Levels of moisture within the rooms are slightly above levels that were quoted as normal in British Standards current at the time, but possibly of more consequence is the general level of heating. This was well below that recommended by the local electricity supplier. Certainly a good level of background heat would enable more moisture to be held by the air and then carried out of the room through natural ventilation.

It appeared that there was insufficient natural ventilation to remove this moisture laden air. Owing to the design of the bungalow, moist air from the bathroom tended to move towards the bedrooms. The kitchen, with its extract fan, was not badly affected, though such effects as heat gain and convection currents set up by the deep freeze and refrigerator would help to reduce the problem.

Remedial treatment

Adequate heating and moisture removal at source are necessary to prevent condensation on the walls at ceiling level. An extract fan in the bathroom will help. This, and the kitchen fan, should be left running for a time after moisture-producing activities have ceased.

Brief flow visualisation tests showed that air movement caused by the direct electric heating systems was minimal and air change rate was low. Any solution must ensure that ventilation is not excessive, otherwise draughts would cause discomfort and expensive heat loss. With this type of heating, warm-up of the air is rapid, resulting in air temperature levels that are acceptable whilst the building fabric is still cold; under these conditions condensation can rapidly become a problem. People rarely use ceiling heating on its own and usually achieve a better level of heating by topping up with another type, such as an electric panel heater or small fan heater.

If the wall cavities are filled with insulation, the temperature of the walls would improve only slightly; on its own, insulation would not cure the condensation problem. However, the benefits gained in heat saved should help to offset the additional heat required for increased ventilation.

CASE STUDY — Water dripping from the ceiling in a fine-art store-room

Condensation had been a problem in the picture store extension of this building since its completion, so much that the room could not be used. BRE was asked to inspect the extension and suggest remedial measures.

Construction and history
The new construction was a first floor extension running the width of the building which was approximately 25 m at the north end.

The walls were generally of solid 337 mm brickwork with some panels of 300 mm cavity brickwork. A course of soldier bricks extended round the perimeter of the building about 1 m below roof level. Above this level the wall was solid 225 mm brickwork faced with Darley Dale stone about 100 mm thick. Just below roof level, a continuous massive stone cornice projected about 225 mm beyond the face of the building. The parapet wall, above the flat roof, was 112 mm brickwork faced with 100 mm of stone. A weathered stone coping approximately 400 x 150 mm on a lead core DPC topped the parapet wall.

The flat roof consisted of 100 mm precast concrete plank units with foamed cores, the units spanning between 400 x 150 RSIs. Black sheathing felt had been applied to the planks before laying a lightweight screed, which was in turn finished with 20 mm asphalt. The lightweight screed had been laid to falls; generally the screed was at its thinnest at the perimeter walls. The roof was pierced by the large double-glazed gable-end roof lights carried on cast in situ concrete kerbs.

The extension could not be used as a picture store because of drips from the ceiling principally over triangular areas (sides approximately 7 m) at the north-east and

south-east (exposed) corner of the room. Condensation had also occurred in the roof lights and work had been undertaken to cure this problem.

Having failed to find leaks in the roof some years before the BRE visit, the owner considered that the problem was most likely to be condensation. It was therefore decided to improve the thermal insulation of the roof so a suspended ceiling of 25 mm thick insulation panels with specially taped joints was installed. No further problems were experienced until the onset of cold weather the following winter. At this time, and subsequently, the north-east and south-east corners of the room were again troubled with drips of water, this time running down the walls. Ceiling panels in the affected areas had been removed for inspection, and beads and runs of water had been observed on the underside of the concrete planks. The panels had subsequently been replaced and retaped.

The condition of the air within the storage room was controlled at 20.5°C and 56% RH, dewpoint 11.7°C, vapour pressure 13.8 millibars: a mixing ratio of about 8.6 gm/kg dry air.

Site observations
The weather on the day of the visit was damp and cold, with conditions improving slightly during the day. At 15.30, outside air conditions were dry-bulb 10°C, wet-bulb 8.6°C, RH 83%, vapour pressure 10.2 millibars.

Using an aspirated hygrometer, air conditions were measured at 11.00 at eight positions round the room at about 1.5 m above floor level.

Dry bulb °C	Wet bulb °C	RH %	Dewpoint °C
19.5	14.6	58	11.3
20.1	14.7	55	11.2
By installed thermohygrograph			
20.5	15.0	54	11.4
20.1	14.8	56	11.4
20.1	14.8	56	11.4
19.4	14.5	58	11.3
At ceiling level			
19.9	14.6	56	11.1
Hole through suspended ceiling into roof void			
15.2	12.0	68	9.8

REFERENCES AND FURTHER READING IN CHAPTER 4

Robust details: Limiting thermal bridging and air leakage - robust construction details for dwellings and similar buildings. London, The Stationery Office.

Building Regulations England and Wales. Approved Document L. Conservation of fuel and power. London, The Stationery Office.

English house condition survey 1996. Department of the Environment. London, The Stationery Office, 1998.

Scottish house condition survey 1996. Scottish Homes. Survey report. Edinburgh, Scottish Homes, 1997.

MATCH software.
http://www.match-box.dk/dk/hvaderbeskrivelse.htm

Energy Efficiency Best Practice programme (EEBPp) Good Practice Guides
174 Minimising thermal bridging in new dwellings
183 Minimising thermal bridging when upgrading existing housing

BRE publications
Roofs and roofing BR302

Walls, windows and doors BR352

Floors and flooring BR460

British Standards Institution
BS 5250: 2002 Code of practice for control of condensation in buildings

82

Chapter 5
Rain penetration

Driving rain and the driving rain index
Rain penetration in walls
Rain penetration at openings
Rain penetration in roofs

This chapter tells you how to assess the risk of specific designs in actual locations in the UK using the driving rain index, and deals with rain penetration in solid and cavity masonry walls, cavity wall insulation, cladding systems, DPC detailing principles and well-tried details, rain penetration of pitched and flat roofs, parapets and leaking windows.

Figure 5.1 A disfiguring deposit of carbonate from rain penetrating the sloping brickwork parapet

Figure 5.2 Although much of this results from condensation, there is also some rain penetration

There are regional construction differences throughout the UK as a result of local experience and practice as well as available materials. In more exposed locations, walls may be sand:cement rendered and slate or tile-hung in Cornwall and Scotland. Pitched roofs are given a second line of defence with a sarking material of felt or plastics. In Scotland, boarding is used as the sarking. Windows in Scotland are usually inset to give protection; other parts of the UK use a narrow sill with the window much closer to the line of the outer leaf.

DRIVING RAIN

In the Building Regulations, control of moisture is a functional requirement and the building must be designed to adequately resist such penetration – see Approved Document Part C and Part G.

An International Council for Research and Innovation in Building and Construction (CIB) *Working Commission on Rain Penetration* meeting in the 1950s adopted a definition of rain penetration:
By rain penetration is meant that rainwater penetrates into a wall either through the surface of the wall, or due to leakage at windows or similar installations. It is not necessary that water penetrates so far that it may be discernible on the inside of the wall. More information is in *Rain Penetration Investigations - A summary of the findings of CIB Working Commission on Rain Penetration - Oslo 1963.*

Rain penetration in modern cavity walls tends to show as a well-defined

roughly circular area on internal finishes. Sometimes surface salts will define the outer limits of such wetting. If the wetting persists, most of the wall may become visibly damp. In older, solid wall buildings, wetting may not be visible because successive coats of emulsion paint or vinyl wallpaper have masked the effects. The extent of the dampness, or if dry the salts which define it, can be traced with a moisture meter.

Moisture can be deposited on external surfaces in several ways:
❑ Gentle rain or drizzle normally falls vertically and will accumulate on flat surfaces. Some splashing may wet adjacent surfaces.
❑ Driving rain, which is heavy rain blown by a strong wind on to horizontal and vertical surfaces. Water can also be blown uphill on sloping surfaces.
❑ Snowfall and wind-blown snowdrifts have little effect at the time but when the snow melts, it can cause severe wetting, particularly very fine snow blown into pitched roofs.
❑ Fog wets external surfaces but in small quantities and has little effect.
❑ Condensation can occur on outside surfaces in tropical climates, particularly with air-conditioned buildings. Storms in these climates are more likely to be a test of weathertightness.

Figure 5.3 Severe wetting from driving rain on an exposed wall

Figure 5.35 Using an optical probe. Inset is a typical field of view with a camera connected to the probe. The wire tie in the cavity is relatively uncontaminated with mortar droppings

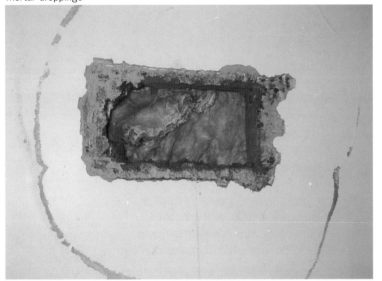

Figure 5.36 Salts have dried and destroyed the paint film. Removal of the block revealed a corner of the bat turned over

Signs of rain penetration

☐ Damp patches following heavy rain. If the patches appear immediately, the cause is probably hygroscopic salts, often seen on chimney breasts or where there has been local salt contamination.

☐ Patches on the inside of external walls exposed to the prevailing weather, usually showing efflorescence. They can be related to faults in the wall itself or to defective rainwater goods or parapets.

☐ Dampness round window and door openings is usually due to incorrect installation of DPCs and cavity trays.

☐ In ceilings, damp patches usually point to a leaky roof or faulty guttering.

Survey methods

Finding out exactly what kind of wall construction has been used in a building is often not easy, and some destruction in the interests of diagnosis of a problem may be inevitable. It will be a matter for professional judgement in balancing the severity of problem with the consequences and costs of making good any damage.

Optical probes may be useful because they keep damage to a minimum but our experience is that they must be used with caution, since the field of view is restricted – Figure 5.35.

Diagnosis

Before taking any remedial action against rain penetration, exclude all other possible sources of dampness. Visible effects often dry out fairly rapidly although repeated wetting will mark decorations and may leave a deposit of salts at the furthest limit of the damp – Figure 5.36.

Outlining and dating the extent of wetting is a useful reminder of what has been experienced. If circular patches have occurred in a fully filled cavity wall, dirty wall ties may be the problem. However, opening up the wall may not be conclusive and drainage of water through poorly installed fill materials is more difficult to quantify. A spray test using a sparge pipe is useful in confirming rain penetration and subsequent opening up of the cavity should help to identify the way the water is reaching the area of dampness. More guidance on diagnostic methods is given in *Rain penetration through masonry walls: diagnosis and remedial measures*.

The main symptoms of rain penetration through walls are summarised in the box below.

If rain penetration is correctly diagnosed as the problem, it can be difficult to pinpoint the exact route the water is taking. A damp patch on a ceiling could be due to a missing tile some distance away; masonry in parapets and chimneys can collect rainwater and deliver it to other parts of the building below roof level; blocked gutters can lead to damp patches on walls that appear to be rain penetration.

Remedies: external

Of the many recommended treatments for walls that have become damp from direct rain penetration, some are applied to the outside, some to the inside surface and some can be used on either surface. The advantage with external treatments is that they keep the body of the wall dry, and so help to improve both durability and warmth.

Figure 5.37 Tile hanging in progress. Unfortunately the 'breather' membrane is not a breather membrane at all but is a polyethylene sheet. This forms a vapour control layer on the cold side of the thermal insulation with a risk of condensation inside the polyethylene

Tile or slate hanging

The most effective of all external treatments is tile or slate hanging. It is durable, especially if any timber used in fixing is pressure-impregnated with preservative. Tile or slate hanging has an advantage over most other treatments in that it is able to shed most of the rain failing on it without impeding evaporation of any moisture that may still find its way into the body of the wall by indirect paths.

Provided tiling or slating is constructed without obvious faults, the wall should be weathertight for all exposures of driving rain, although wind may lift tiles or slates, as it can with roofing. A suitable breather membrane to BS 4016 – Figure 5.37 – is important; more information is in *Roofs and roofing*.

Renderings

External renderings in cement, lime and sand mixes are next in effectiveness to tile hanging. They are particularly useful in preventing direct penetration through cracks between mortar and bricks or blocks. Applied to a nine-inch one-brick wall, they can be expected to keep the internal surface of the wall dry except when it is exposed to severe driving rain. This may also occur towards the end of an abnormally long spell of wet weather when there has been little opportunity for evaporation.

A remedial measure for rain penetration in a two-storey house is to provide a render to the first floor level only (including the gable), with a bellmouth finish at the lower edge. This is often cosmetically more acceptable, and cheaper, than complete cladding or render. The mixes usually recommended for render are cement:lime:sand 1:2:9 or 1:1:6 These produce finishes that are porous, absorb water in wet weather and permit free evaporation when the weather improves; the action is rather like that of a thick layer of blotting paper. Full details for the choice of mixes are given in Good Building Guide 18 *Choosing external rendering*.

A dense, impervious rendering might seem preferable but is often less efficient than a porous one. To be effective, dense rendering must be free from cracks, a condition difficult to ensure. If cracks form, rain water running down the face drains through the cracks into the body of the wall and becomes entrapped behind the dense rendering; this impedes subsequent evaporation. Consequently, the moisture travels inwards, much of it evaporating from the inner surface, causing familiar signs of dampness, such as efflorescence and staining. Penetration through cracks in a dense rendering is most marked where the rendering has a smooth surface. Such finishes as roughcast and pebbledash shed much of the water that falls on them and are less likely to cause dampness in this way; they are particularly suitable where the exposure is severe, for example near the coast.

Renderings are often used to increase the weather resistance of a basic wall material; however, the more exposed the wall, the more restricted is the choice of render. The factors involved are:

- Rendering can increase the exposure rating to wind-driven rain of a wall by one or two categories: a wall which would otherwise be suitable only for sheltered exposures can be upgraded to severe or very severe – see *Thermal insulation: avoiding risks*.
- Exposure to marine or polluted environments may lead to attack on the cement content of renders or increase the rate of corrosion of metal lathing.
- Rendered walls exposed to driving rain and pollution may streak differentially. Flint or calcareous dry dashes have the best self-cleaning properties though some flints contain iron, which can lead to staining.

Renders have a significant effect on reducing rain penetration into walls but their effectiveness may be variable:

- 1:1:6 and 1:$^1/_2$:4$^1/_2$ renderings are effective in reducing the passage of water into brick backgrounds and this is improved further by adding a dry-dash finish. Performance can be significantly reduced by cracking or loss of dash by erosion.
- Rendering does not reduce the passage of water into aerated concrete backgrounds as much as it does on clay brick backgrounds. It does, however, help to prevent rain penetration through the joints.
- Evaporation rates are generally much lower than absorption rates but there is usually no significant build-up of water within the materials.
- Rainwater absorbed intermittently is usually lost by evaporation before it can penetrate deeply.

Timber boardings and PVC-U sidings

Timber feather-edged shiplap is comparatively weathertight provided it does not rot, warp nor split. Traditionally, the timber is painted or treated both for appearance and for protection. PVC-U sidings are unlikely to suffer too much from these if they are not vandalised but there can be surface deterioration. Traditional paints can reduce impact resistance so special products have been formulated – see Digest 440.

Vertical boarding spaced apart to allow ventilation is frequently seen on agricultural buildings. Provided the joints are not too wide, little driving rain will enter the building, especially when it comes at an angle inclined to the wall. Wind-driven fine dry snow, however, is another matter.

Paints and other coatings

Paints have been used on stucco for centuries, generally successfully. The main problem is that the paint film may crack and allow water to wet the masonry or render behind. Drying is inhibited by the paint film and can lead to an accumulation of water and potential frost or sulfate damage. Cement paint coatings are particularly good at shedding water under severe exposure. Weather resistance is improved only if the wall has no major cracks or defects that allow water to penetrate behind the paint film.

Aerated concrete blocks or panels must be coated to resist rain penetration. Investigations by BRE on different coatings for water penetration, vapour resistance and durability demonstrated the significant influence of coatings on moisture content gradients across the wall. There were persistent high levels or even penetration to the inner surface under some conditions,.

Oil, bitumen and tar paints were traditionally used to coat the plinth at the base of the wall. They give an almost impervious surface coating, so they should be used only on walls where there is no risk of indirect penetration of moisture through parapets, sills, etc, nor any likelihood of a build-up of condensed moisture at the back of the paint film.

Solidly bedded tiling

As with any comparatively impervious finish which is prone to cracking, the weathertightness of a wall covered with solidly bedded tiles depends on the absence of cracks which allow rain water run-off to penetrate the finish. It may not easily find a way out, and frost action can then occur.

Colourless waterproofers and repellents

These treatments can improve weather resistance. They are clear and at worst will give only a slight sheen to the wall. Although water-repellent treatments do not completely seal the pores of the surface, some closing of the surface is inherent. The principle is that a greater quantity of water is turned away from the surface than is prevented from evaporating. Run-off from the surface will be increased so vulnerable details which might leak are at greater risk.

BS 6477 describes test methods for water-repellent treatments but they do not take into account either the crucial influence of mortar joints or the effect of water applied with forces similar to wind-driven rain. BRE tests on clay and calcium silicate brickwork using BS 6477 methods showed silicone-based treatments to be the most effective; treated masonry did not leak and the moisture content of the walling remained low. Polyoxoaluminium stearate treatments were moderately effective but one based on acrylic polymers was poor.

Colourless waterproofers make a wall surface water repellent and less porous, without much change in appearance. Like paints, they should be used with discretion; in particular, make sure that the dampness is not due wholly or in part to some cause other than direct penetration through the wall. If it is, the attempted cure may cause more trouble. The permanence of the protection is variable and depends on the type of waterproofer and on the condition of exposure. Periodic renewal will probably be necessary.

Repointing

Repointing should be the only maintenance required on a durable brick; its frequency depends on the mortar, the finish of the joint, and the degree of exposure of the wall; a life of at least 30 or 40 years ought to be expected. Hand-raking the old mortar wherever possible is preferred to mechanical equipment. Disc cutters can cause considerable damage.

Leaking mortar joints should be deeply raked out and repointed using a mix compatible with the existing mortar and type of masonry. It can be difficult on site to sort out the various designations of mortars, so there is something to be said for the use of a 'general use' mortar, which can resist all but the most severe exposure, and can accommodate minor movements. Such a mix is $1:1:5^{1}/_{2}$ Portland Cement: hydrated lime:Type S or G sand plus an air-entraining agent.

Pointing technique can play havoc with durability and can alter the character of a wall. Recessed joints, for example, can lead to reduced weathertightness so should be used only in the most sheltered locations.

Remedies: internal

If you cannot be certain that direct penetration of rain is the sole cause of dampness, internal treatments have obvious merits. They may help to combat dampness due to rising ground moisture, indirect rain penetration or contamination with deliquescent salts. A further advantage is that they are applied to a more accessible surface. External treatments could aggravate this sort of dampness, unless they are porous.

Dry lining

The traditional way of preventing moisture reaching interior wall surfaces was to batten out (strap) and fix wallboard, t and g or lathing out of direct contact with the wall. All timber used for plugs and battens was pressure-impregnated with a non-staining inodorous preservative. Where they are in contact with plugs or wall, battens should be painted with a bituminous paint to reduce transmission of moisture, or a slip of bitumen felt should be inserted between the wall and batten. If possible, the space behind the lining should be ventilated.

An alternative was to fix corrugated bitumen impregnated sheet to the wall and render or fix plasterboard over. This has now been replaced by plastics dimpled sheets with the options of similar finishes. Both methods need ventilation behind the impermeable sheet. They take up an appreciable amount of space and sometimes present difficulties at openings.

Dense internal renderings

These were commonly used in the 1920s and 1930s with solid brick walls and can be equally suitable as a remedial treatment for damp walls. A dense cement and sand rendering, often with an integral waterproofer added, is used; such a rendering impedes the passage of moisture to the inner surface but it also slows down the rate of deterioration of the decorative coating. It does not prevent all penetration but can, in favourable circumstances, reduce it to an acceptable level.

The suitability of an internal rendering depends largely on an alternative escape for any water that enters the wall. This water must be allowed to evaporate elsewhere. If the body of the wall and any external covering is a porous material, the internal rendering is likely to have a significant effect. If, on the other hand, evaporation from the outside surface is likely to be difficult, little benefit can be expected. This can be because the wall itself is dense or there is a dense rendering on the outside.

The value of dense renderings used in this way depends to some extent upon the nature of the decoration. Finishing coats of calcium sulfate plaster and decorative coatings of oil paint, distemper and wallpaper are all sensitive to dampness. Where circumstances permit, there is much to be said for omitting the finishing coat plaster and applying cement paint directly on the cement rendering.

Internal waterproofers

Another internal treatment for damp walls is to apply an impervious coating of bitumen or similar material; this is followed by blinding with sand and plastering or by lining with wallboards. The adhesion of the impervious coating to the wall is critical and can be lost if the wall is wet when the waterproofing is applied. This treatment can be successful but BRE's experience is insufficient to assess the risk of failure in any given circumstances.

It is sometimes thought, mistakenly, that 'sealers' applied to a damp wall will prevent the dampness reaching the inner surface and spoiling the decoration. The principal function of most proprietary sealers is to reduce suction and so facilitate application of a decorative coating. No doubt they reduce the porosity of the surface plaster, but they cannot be expected to make it waterproof.

The same is true of alkali-resistant primers. They guard against attack on a paint film by alkalis in the surface of the wall. They will not prevent other effects of dampness, such as blistering, loss of adhesion and efflorescence.

Workmanship

Unfilled and partially-filled cavity walls
The precise location of the dampness must be pinpointed before any defect can be remedied. You may have to remove areas of paint or wallpaper and use a moisture meter to locate the boundaries of damp patches. A wetting test will confirm that the dampness is due to rain penetration.

Pinpoint the location, then remove a few bricks from the inner or outer leaf and inspect the cavity using a mirror and a torch with a narrow beam. Alternatively, use an optical instrument such as a borescope.

You can remove small obstructions in the cavity, such as mortar droppings on ties, from holes drilled in the inner or outer leaf. Other faults, such as misplaced DPCs, may also be diagnosed by looking into the cavity.

Some types of metal detector are useful for finding the position of wall ties. If the position of a tie corresponds to a damp patch, it is worth removing a brick or block at that point for detailed inspection.

All these methods are possible for unfilled walls and for those with partial cavity fill. For more details, see *Rain penetration through masonry walls*.

Fully-filled cavity walls
It is quite difficult to identify the cause and location of rain penetration in a fully-filled wall; a special investigation may be needed.

A wetting test is necessary for some fills; simply opening up the cavity without artificial wetting can be deceptive. Penetration through built-in fibre batts usually occurs at joints between batts rather than through the batts themselves. Mortar bridges or other minor installation faults can be corrected quite easily, and the joints re-formed before closing up. For injected fills, injection of more material sometimes cures the problem.

Repointing can help if mortar joints are cracked or in poor condition.

In exposed areas, or if the dampness is widespread, one of the remedies for solid walls described above is probably best.

Renderings
It is extremely difficult to repair a cracked dense rendering satisfactorily. Even if all precautions are taken to undercut a groove, wide enough to fill on the line of the crack, and then to fill it carefully with a cement:sand or a cement:lime:sand mix, a new crack will probably form at the side of the filling within a year or so. Although narrow at first, the new crack might be filled satisfactorily by a cement paint treatment but, sooner or later, the

major repair of cutting out and filling with mortar will be needed again. There is much to be said for making a thorough job of the repair by removing the whole of the dense rendering and replacing it with a less dense mix of cement:lime:sand. This more porous rendering is less likely to develop large cracks and should not cause entrapment of moisture within the wall.

Panelled walls

There is no strong case for an overall treatment of a wall of this nature where the joints, rather than the panels, have allowed rain penetration. Filling the joints with a suitable mastic, which can accommodate small movements, will often prevent further moisture penetration. The mastic will of course, need periodic renewal.

Assessing repairs

Make adequate allowance for the time required for heavy masonry construction and concrete to dry out; a year may be too short a period if the material itself is very dense (good quality concrete for example) or if the wall is covered with an impervious skin of tiles, dense rendering or paint. Efflorescence may continue to appear on the surface during the whole of the drying out period

There may also be deliquescent salts present in the damp wall; while it is drying, they will be brought to the surface, possibly in sufficient quantities to cause damp patches. Further local treatment will then be necessary.

Damp-proof courses

Purpose and performance

Damp-proof courses (DPCs) are formed from impervious materials introduced to prevent the passage of liquid water from one part of the construction to another. This usually means attempting to keep the inner skin free from absorbing moisture and therefore 'dry'. Chemical DPCs are used specifically for remedial work at the base of the wall. DPCs in solid walls are normally found at the base of the wall and in parapets. One example was formed from lead and was through the full width of a parapet wall. But there was an inherent defect: the flashing was on top of the DPC and the lead was perforated and required replacement – Figure 5.38 and page 146.

Requirements for DPCs become more critical within the cavity wall. The risk of failure is high unless the concepts are understood and material selection and workmanship carefully approached. BRE Advisory Service has been involved in many hundreds of cases of DPC failure and the remedial work can be costly – Figure 5.39. Demolition of whole sites has been necessary in some cases. The introduction of pre-formed components has reduced the incidence of failure but has not eliminated it.

Figure 5.38

Figure 5.39

Location

Figure 5.40 shows situations where DPCs are required and where they can be omitted.

A DPC is needed:
- ❑ in both leaves of external cavity walls to prevent rising damp; it should be positioned at least 150 mm above adjacent ground levels to guard against rainwater splashing the wall from the ground – Fig 5.41
- ❑ beneath internal partition walls; this prevents a wet slab drying out into the base of the wall;
- ❑ beneath sills and copings which are formed of jointed units to prevent penetration of water to the wall below – Figure 5.42.

A DPC should be installed in a parapet wall to provide continuity of the weatherproofing with the roof covering. It should be not less than 150 mm above the roof finish to lap over any cover flashing to the roof upstand. Some designers feel that taking the tray outwards leads to unsightly staining running down the outer face of the masonry. There is the option to drain the tray towards the roof but this concentrates water at a vulnerable junction. If the cavity is filled, the tray should always drain to the outer leaf.

Where a masonry chimney penetrates a roof structure, one or two DPC trays with appropriate flashings should be provided. If the capping is jointed, a DPC is required below – Figures 5.43 and 5.102.

Openings in cavity walls need protection with integrated vertical and horizontal DPCs to deflect water away from the inner leaf – Figure 5.44. Vertical DPCs are needed at the jambs of openings where the cavity is closed; horizontal DPCs are also needed at the head to bridge the cavity in the form of a tray. A DPC is necessary where the sill is jointed.

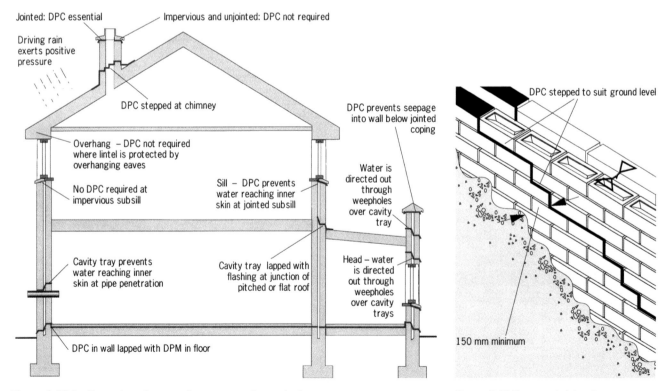

Figure 5.40 Positions where damp-proof courses may be required

Figure 5.41 To prevent rising damp

The appearance of plain tile verges is enhanced by tilting the verge tiles inwards. A small amount of tilting does not affect performance. Single-lap interlocking tiles should not be used in tilted verges because their performance will be affected.

Dry-laid verges

Dry-laid verges are an alternative to ordinary verges; the specially shaped tiles form a downstand similar to a bargeboard. Potentially this form of verge will be more weatherproof than the mortared-and-pointed verge, provided the fixings are secure.

Bargeboarded verges

A bargeboarded verge gives more protection to the wall than a verge set on the wall head – Figure 5.73 – but ensure that:

❐ the underfelt is carried over the top of the bargeboard to meet with the undercloak;
❐ the inboard edge of the ladder is securely fixed to the rest of the roof;
❐ the cantilever effect of the tiles does not impose too great a load on the bargeboard.

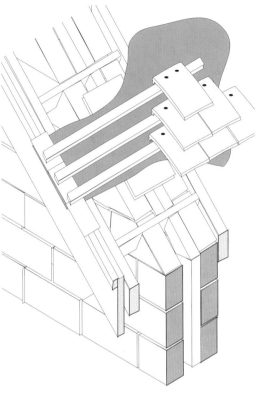

Figure 5.73 A bargeboarded verge with gable ladder

A verge detail common on pantile roofs in Suffolk is a section of timber nailed to the top of the bargeboard, overlapping the verge tiles – Figure 5.74. This serves the dual function of providing a reasonably weatherproof side lap for the verge tile and a positive fixing to resist suction in that part of the roof which is most sensitive to wind. Durability is the main problem because the section can be painted underneath only by removing it.

Abutments

Where a lower roof abuts a higher one, an external wall becomes internal below the roof – Figure 5.75. A completely impervious sheet material must be provided because masonry alone is inappropriate. Lead was traditional and especially effective where it was protected by bituminous paint; many other materials, such as copper and bituminous felt, are satisfactory, or sandwiches of thin sheets of these materials. Other materials are described in BRE Digest 380.

Figure 5.74 Common verge detail in a Suffolk pantile roof. The timber cappings shown are prone to decay since it is impossible to repaint the underside without removal

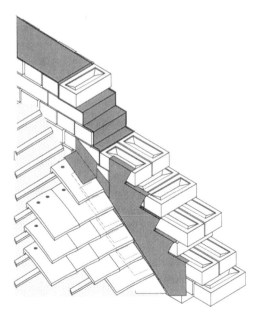

Figure 5.75 A DPC, cavity tray and flashing is needed to prevent rainwater percolating down a wall: not easy when the pitch of the roof differs from the rake of the bond

Sarkings and underlays

Rain and snow can be forced past the edges of lapped tiles and through the joints by wind. In theory, at least, tiled and slated roofs with boarded sarking, as is common in Scotland, will be at less risk of rain penetration than roofs having flexible sheet sarkings since the boarded sarking performs much better as a wind barrier. Sarking membranes should have a vapour permeability in the range 0.1 – 2.0 MNs/g.

On one-third of building sites visited by BRE, the installation of sarking felt underlays was faulty in various ways. In particular, sarking felts were not fitted closely around soil and vent pipes, nor properly lapped nor dressed out to eaves gutters and bargeboards.

If the space underneath a layer of tiles is not sealed, there is a risk of fine snow being blown between the joints, even though water may not penetrate. The current Scottish boarded sarked roof can be expected to perform much better than the English counterpart. The unsupported underfelt used in a traditional English roof can billow in the fluctuating wind pressures of a snow storm, allowing the snow to be pumped through the gaps in the covering.

Gutters and rainwater pipes

Although thatched roofs that have a wide overhang are not normally fitted with gutters, most other roofs have a system of gutters to catch the water running off the roof. Downpipes then lead the water for disposal. At the extremity of the eaves of a pitched roof, the shape characteristics of the covering have an influence on the positioning of the rainwater gutter. The ideal profile of the covering (from the point of view of rainwater disposal at the eaves) is for the upper edge to be rounded and the lower edge to be sharp to provide a drip. The raindrops will then have less chance of being blown back up the underside of the slope. The minimum projection into the eaves gutter is about 50 mm. BS 6367 shows the normal trajectory of droplets from drips of various profiles. Further protection to the eaves can be given by overlapping the lower edge of the sarking felt (or a substitute, a more durable felt strip) into the gutter.

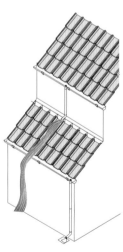

Gutters are not usually designed for storm conditions and maintenance is frequently neglected; the system is ineffective if outlets are blocked with leaves. The worst cases are valley gutters where the discharge of the increased water load from large areas of roof completely overspills the eaves gutters, soaking the wall beneath.

It is common for rainwater pipes to discharge over lower roof slopes. On a roof of plain tiles or slates there is usually no problem since the discharge fans out before reaching the lower gutter – Figure 5.76 – but this is not so with heavily profiled interlocking tiles. The profile confines the whole of the run-off to a single valley and there is risk that the tile laps will leak. The discharge at the foot will almost certainly overshoot the lower gutter in all but the lightest rainfall. The tiles will also be preferentially stained.

Figure 5.76 Patterns of discharge from rainwater pipes over plain and profiled tile roofs

Proprietary precast concrete gutters often have been fitted at the head of walls where the roof is of fully supported felt construction; in the past they have also been used on pitched roofs covered in different materials, and as permanent shuttering for lintels carrying the eaves and wall plates over first-floor windows. Originally lined with bitumen felt, they were, in our experience, a continual source of leakage caused by splitting of the linings which were fully bonded to the concrete, the joints of which had opened.

Flashings

Loose flashings allow rain to penetrate behind; especially at abutments, flashings may need additional clipping. We have seen an example where, on a severely exposed site, the contractor had applied a silicone sealant between the brickwork and the lead in an attempt to keep water out – Figure 5.77.

Flashings should be chased a minimum of 25 mm into brickwork. Particularly vulnerable is the change of slope in a mansard where the flashing and the first row of tiles above it can be stripped by wind action. The heavier the weight of flashing, the more it can resist wind action; the lower edges of thinner flashings in this position should be held down with strip tacks. Code 5 is more appropriate than Code 4 for lead flashing without tacks. Poor detailing of sarkings is common with tears in sarkings leading to problems with rain or melting snow.

Figure 5.77 A silicone sealant between the brickwork and the lead in an attempt to keep water out.

Valleys

About two-thirds of roofs of new houses built between 1991 and 1993 had valley gutters. Before this most valleys were formed by inserting a metal lining into the valley and cutting the tiles to shape over. There was much clumsy cutting, with appearance compromised. Better solutions, though seen more rarely, used purpose-made valley tiles of various configurations. Valleys formed from plain tiles may be swept or laced, either to a radius or to an acute angle – Figure 5.78.

Figure 5.78 A swept valley in plain tiles. The camera viewpoint has foreshortened the tiles, exaggerating their apparent irregularity. The parapet capping has not succeeded in throwing rainwater run-off clear of the wall

A double thickness of sarking felt is needed in valleys, lapped at least 600 mm over the centreline of the valley. Tiling battens must be properly supported where they abut valleys.

Our experience is that the design of valley gutters is frequently defective. We have seen drawings that specifically state that the design of gutter systems should be left to the site staff to sort out! Valley gutters slope much more shallowly than the pitch of the roofs they join; they are particularly prone to leaking on pitches of less than 20°.

Incidence of defects

In a large sample of new housing in 1992 – 93, we found that almost 20% had unsatisfactory weatherproofing at abutments and that in about 10% the bottom course of interlocking tiles had been tilted upwards at the eaves; the tiles had not interlocked satisfactorily, so making the eaves more vulnerable to rain penetration.

As can be expected, there are many different kinds of weathertightness faults in older roofs. Some of the more common ones we found in older pitched tile roofs include:

❏ absence of sarking and torching;

❏ strips of sarking tacked to battens between rafters, leaving rafters unprotected;

❏ many cement fillets cracked and displaced;

❏ flashings working out of joints;

❏ clay roof tiles delaminating;

❏ replacement tiles overhanging eaves gutters too far, so rainwater run-off overshoots;

❏ rainwater from long valley gutters overshooting undersized eaves guttering.

Windows set vertically within the slopes of mansard roofs sometimes present problems. The vulnerable point is where protrusion changes to inset – Figure 5.79.

Problems are common in late Victorian terraced houses where local byelaws required that separating walls were built to project through the roofs. Many houses with this feature can now be found 'flashed' with cement fillets or flaunching. There is no real substitute for a metal flashing over a secret gutter, or soakers, but if a cement flaunching is preferred, it should include a gritty fine aggregate with some lime in the mix.

The copings of projecting separating walls for the most part rest on kneelers on the slope or footstones at the eaves, and are bedded in mortar up the slope. Most are in good condition though there are reports of more recent attempts to emulate the detail, usually at gables, without using kneelers or footstones, where the copings have slid down the slope as a result. Damp-proof course materials such as slate should be capable of adhering to both coping and masonry below, or it may be possible to use special fixings. If sheet materials provide a slip plane, there is a risk of the copings dislodging in high winds. Mortars here could have a bonding agent.

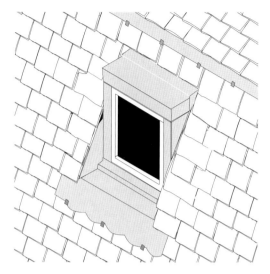

Figure 5.79 Windows set in mansard roofs need careful detailing

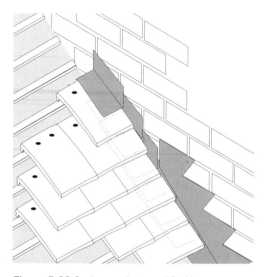

Figure 5.80 Soakers and stepped flashings

Cavity trays, soakers and secret gutters, with associated flashings, can cause problems where there are steps in levels between adjacent buildings – Figure 5.80. Problems seem not so much with design as with execution, but the preformed cavity tray with attached flashings performs satisfactorily.

Where a pitched roof abuts a stepped-and-staggered separating wall, part of the roof becomes an external gable, and the outer leaf of masonry becomes an inner leaf below roof level. Ensure that rainwater is prevented from reaching the interior. Soakers for plain tiles should be 175 mm wide minimum to give an upstand of 75 mm against the wall, and a lap of 100 mm under the tiles. A cover flashing then laps the soakers by at least 50 mm, preferably more. For interlocking tiles, use a stepped flashing dressed over the tiles; soakers cannot be used because they would interfere with the interlock. A secret gutter is feasible. Lead should be of Code 4 thickness for soakers (Code 3 is sometimes specified, but it is less

which take up water lying against the open edge of the welt and transfer it to the inside of the roof. In addition, water vapour in the atmosphere may similarly be transferred to the roof interior and condense on relatively cold surfaces near the outside of the structure. Loosely referred to as 'pumping', this is attributed to two main mechanisms:

❏ external wind pressures and suctions forcing air to and from the inside of the roof void (this tends to occur, given appropriate conditions, where the voids are relatively large);
❏ temperature changes affecting the air inside the void (tending to occur, given appropriate conditions, where the voids are relatively small).

In both cases, the resultant movement of air through the gaps carries with it some liquid water or water vapour from outside to inside which cannot then easily escape if the void is not ventilated. Significant volumes of water can be transported; enough indeed, to wrongly diagnosis the cause as rainwater penetration through holes or other damage to the covering.

Where the problem has been correctly diagnosed, there are effective remedies: for the first mechanism, by providing better air seals at the welts or joints, effectively making them less responsive to external air pressure; for the second mechanism, by better ventilation to the immediate underside of the membrane via special weatherproof vents situated away from the joints. This should remove water vapour without admitting rain. You may need specialist advice.

When specifying fully supported metal roof coverings, ensure that there is some ventilation over the top of the thermal insulation and underneath the outer sheeting on its deck. Otherwise, the sheet can deteriorate through inadvertent rainwater penetration to its underside. It then fails because solar heat expands and expels the trapped air; when it cools during rainfall, a partial vacuum is created which can suck in rainwater over the top of any vulnerable standing seams. Prevent this by fitting small shielded ventilators penetrating the roof covering.

Lead is particularly susceptible to deterioration in this way, other metals less so. Stainless steel and copper sheets are inherently more durable than other metals but there is always the risk of rainwater deterioration of any timber supporting structure.

Metal lined gutters are prone to leak, especially if they are blocked by snow and ice (snow boards are needed). They may also be a source of condensation on the underside of the metal, leading to deterioration in the gutter soles.

Weirs are a common requirement to provide upstands at vulnerable joints across the sheets. They must be checked for signs of lifting after strong winds, especially if of lighter gauge materials.

Patent glazing

The original intention of patent glazing was that it did not involve putty. The preformed strip or cord underneath the glass was not the primary weather seal, merely the inner air seal on a two-stage drained system of jointing. Over the years, this simple principle has to some extent been disregarded and now the weathertightness of some so-called patent glazing seems to depend on adhesion of the seal between bar and glazing. Many types of seals have been used, including tapes, cords and strips in solid and cellular form, and gun-applied mastics. British Standards are available for gun grade materials of certain polymer types though not for tape sealants. BS 5889 is obsolete but European standards are being prepared.

A second line of defence with patent glazing bars is the shape of the channel formed in the section at each side of the bar underneath the seating for the glazing. The route to the outside must be kept open for any rain penetrating the outer cover and running down the channel in the bar.

If the glazing is to be replaced at any time, take care that the deflection characteristics of the bars are appropriate to the kind of glazing to be specified or damage could result.

Snow guards to BS 6367 should be fitted above patent glazing. Rainwater pipes should not discharge over patent glazing – Figure 5.87 – for two reasons:

❐ all the water load will be concentrated on a narrow section of gutter at the foot of the patent glazing with the risk of overshooting in all but the lightest of rainfall;
❐ the glazing will quickly become streaked with dirt.

Flat and low-pitch asphalt and bituminous felt roofs

Theoretically, the materials in built-up roofing provide a continuous and impervious barrier to rainwater and there should be no problems provided the surfaces are not perforated. However, damage has been caused to felted roof surfaces of comparatively low pitch during maintenance, for example when fixing aerials on chimneys.

About half the failures in bituminous felt roofs are due to splits produced by localised movements of the substrate. Rain can then penetrate and cause local damage on the ceiling below and will also, of course, damage the deck material. Splits can usually be repaired but the membrane must be isolated from building movements – more details are in Defect Action Sheet 33.

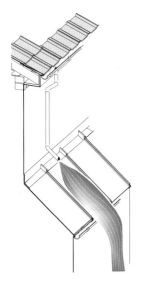

Figure 5.87 Rainwater discharging from a RWP over patent glazing will leave deposits on the glass and overshoot the gutter

The other main cause of rain penetration in flat roofs is poor detailing at abutments and upstands: an angle fillet must be provided where the felt turns through 90°; the upstand must be at least 150 mm high and well lapped by secure flashings. In a parapet cavity wall, the DPC tray must be lapped over the cover flashing.

When BRE surveyed rehabilitated housing, many of the older roofs covered in three-ply built-up felt had problems of rainwater penetration. Many were patched. Roofs using organic fibre felts over boards of high moisture or thermal expansion have not generally been specified since the mid-1970s. Although there were fewer cases of leaks in asphalt roofs, they were invariably found in the oldest of these roofs.

Water penetration of single-ply membranes has been due usually to inadequate joints in the membranes.

Flat roof perimeters

The treatment of the perimeter provides difficulties for the designer; the main options are a parapet, low or higher for safety, if there is access to the roof or just to fit a trim to secure the edge of the membrane. Both have their drawbacks and are likely to be the type of roof detail which is most likely to fail.

Abutments

Abutments, where the roof meets a vertical wall or a parapet, are a common source of problems. The main requirements are for a tilting fillet to turn the roof covering from horizontal to the vertical in 45° increments and well tucked in flashings beneath any cavity trays. The parting of asphalt upstands from their adjoining abutments and parapets is one of the commonest problems with asphalt coverings. The unsupported lead flashing above a valley gutter – Figure 5.88 – was a risky venture; sagging of the lead and not properly sealing the joints encourages early failure. Overflowing of the gutter behind the flashing must be avoided. Fit a weir at the gutter end and ensure that debris never blocks the outlets.

It may be worth installing an automatic leak detection system. One of these works on the principle of a change in electrical conductance in a grid of metal tapes buried beneath, and insulated from, the outer surface of the covering. Water penetration is monitored by a microprocessor so the source can be pinpointed and dealt with.

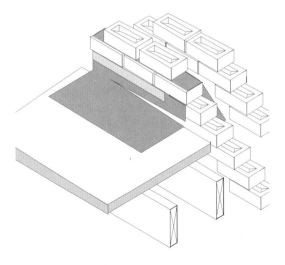

Figure 5.88 Unsupported lead flashing above a valley gutter

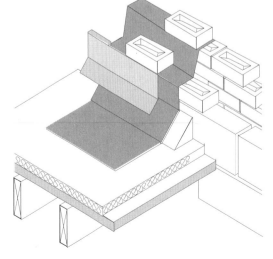

Figure 5.89 A defective abutment detail on a canopy to a doorway. A fillet has been provided at the junction between the canopy and the wall but there is no flashing

Figure 5.90 Defects found at built-up felt abutments (left) and how they can be avoided (right)

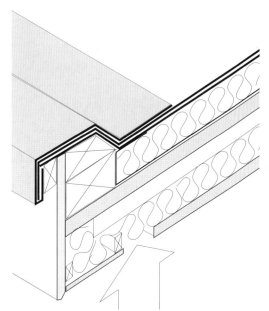

Figure 5.91 Detail of eaves in the timber deck of a warm roof. The external wall in not shown

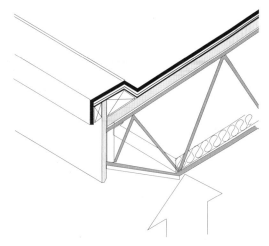

Figure 5.92 Metal trim at eaves of a mastic asphalt cold roof

Figure 5.93 This eaves trim was not securely fixed down to the deck. Consequently outward movement has torn the built-up felt. The shorter the length of metal or plastics, the less will be the thermal movements. Fixing should be made at both ends of lengths, close to joints in the trim

Eaves

The simplest way of finishing the roof edge when using felt is a welted drip at least 50 mm in depth and projection. These welts are normally fixed over a preservative-treated timber batten – Figure 5.91

Mastic asphalt roofs are finished with a metal trim securely fixed as closely as practicable on both sides of each joint in the lengths of trim – Figure 5.92.

Metal or plastics trims are also used with felt or membranes. They must be fixed carefully, or thermally-induced size changes can initiate a tear or crack in the membrane under each joint in the trim, with obvious consequences of a roof leak – Figure 5.93. Water does tend to run off at the junction of the lengths of trim and this should be set forward of the face of the wall to prevent staining.

Asphalt upstands parting from their adjoining abutments and parapets is one of the commonest of problems seen with asphalt coverings.

Parapets

The origin of this detail must be castle battlements where the avoidance of water penetration to the room below was well down on the list of requirements. The damage that can arise from long-term leakage and the consequential high repair costs has already been shown – Figure 5.94.

Rather draconian, but thought worthy of including as a marginal note in BRE Digest 89 was a point made by LB Perkins, speaking at a symposium on Modern Masonry, held in the USA, 1956. He said: "*In our office today, anyone who wishes to put a parapet wall into one of our designs may do so coincident with his resignation*". The guidance in the Digest was to suggest that parapets were to be avoided unless copings had generous overhangs and adequate drips, and were provided with DPCs at roof level. Other recommendations were aimed at minimising the risk of sulfate attack. Attack can cause expansion of the mortar which can arch the coping to a parapet producing wide cracks through which water can obviously enter.

Where a parapet bounds a metal-covered flat roof, melting snow may be a cause of water overflow. The problem arises where ice or snow blocks the gutter so creating a dam; the resultant ponded water overtops the upstands – Figure 5.95. The problem can be solved in part by using snow boards to keep the gutter and outflow clear.

Is it really rising damp?

❑ Are there any leaking or sagging gutters or down-pipes? Clues are green stains on the brickwork, or dampness at low levels on indoor surfaces.

❑ Is there run-off from sills below large glazed areas or below weatherboarding or non-absorbent cladding? If sills don't project enough, they can feed a lot of water into the brickwork below.

❑ Is the building less than five years old? If so, the dampness is probably due to construction water in walls, floors and screeds.

❑ Are there leaks from plumbing, waste pipes or central heating systems, including pipes buried in the wall?

❑ Is it condensation? Persistent condensation on the perimeter of solid floors can wick up into the plaster and produce a stain that looks like a rising damp tidemark.

❑ Has there been any recent flooding, or leaks from a washing machine or dishwasher?

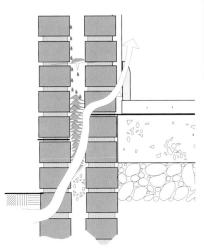

Figure 6.10 Common causes of rising damp

Figure 6.11 Vandals dropped this piece of scaffold board down the cavity

Accumulation of dropped building mortar at the base of the wall cavity can cause bridging. As with solid walls, previously installed remedial DPC systems may be ineffective, particularly osmosis types, those relying on evaporative tubes, and poorly installed chemical systems.

Materials for DPMs and DPCs

Since most forms of masonry allow the passage of moisture upwards from wet ground, some form of protection is needed against rising damp. The moisture barrier usually consists of a membrane laid at a suitable height above the splash zone, and linked with the horizontal DPM. The ideal material for a DPC would be completely impervious to water in both liquid and vapour forms. In practice, not all materials have been effective.

Two courses of slates laid in cement mortars to break joint, and two or more courses of engineering or 'blue' bricks, have been widely used in the past, especially in one-brick solid walls; in many cases they have been effective, especially where the vertical joints were left unfilled. However, the materials in the units themselves are more resistant to rising damp than the mortar used to joint them, and it has become customary (though arguably unnecessary) to make sure that the damp-proofing is still effective by adding a replacement.

All materials should be able to accommodate slight movements in the wall. Sheet materials must be carefully jointed, usually achieved by lapping at least 100 mm. The weight of the masonry wall is usually sufficient to seal the lap without additional jointing materials. The organic felts in some DPCs may have perished but this does not necessarily mean that they become ineffective provided they are not disturbed by subsequent movements.

Bituminous felt DPCs can be squeezed out slightly under pressure, especially in hot weather, but the amounts exuded are usually insufficient to compromise their performance and durability. They are relatively ineffective against horizontal displacement, in effect providing a slip plane on which the walling can move quite easily. Polyethylene is common now but plain sheet has low shear bond; the newest types have a moulded pattern to improve shear and flexural bond. Higher performance DPCs are often formulated from a blend of pitch and polymer; they have good resistance to squeezing.

New build
Building Regulations require a DPC at the base of walls. In cavity wall construction on a concrete slab, common practice is to use the inner leaf as permanent shuttering and place the DPC on top of this level of masonry. The screed has to be isolated by a vertical DPM otherwise drying of the screed into the wall may result in rising damp. The DPC in the outer leaf must be at least 150 mm above the outside ground level. If the site slopes, the DPC must be stepped. There are special arrangements where radon is an issue.

Treatment
Treatment for rising damp involves either curing the source of the problem or masking its effects. Curing at source is normally preferable and, as a first step, the building should be inspected closely to ensure that an existing DPC has not been bridged. Remove earth that has piled against external walls and cut back external render and internal plaster to just above the DPC line. Improving the land drainage at the base of the wall may also help. In old houses with suspended wooden floors, the DPC usually runs in the mortar course at the top or bottom of the ventilation grilles set in the external walls. Adequate sub-floor ventilation is most important so clear the grilles of any obstruction. Check suspended wooden floors for wet or dry rot. They may need extensive preservative treatment and cutting out of any decayed timber. In many of these situations, the original DPC will have been completely effective, a fact often overlooked when considering treatment.

The choice of DPCs
The methods for installing a new DPC are described generally as 'traditional' (the insertion of a physical DPC) or 'non-traditional'. We strongly recommend that you consider non-traditional methods only if they have been awarded an Agrément or other third-party certificate. Chemical injection is the only method that currently satisfies this requirement and is the only method which BRE considers suitable where it is not possible to insert a physical DPC. Physical DPCs can only be placed in brickwork or coursed stonework; random flint walls or rubble infilled walls are not suitable. Unusually thick walls can rarely be treated and it can be dangerous to attempt installations of this type for structural reasons if the walls have settlement cracks. Chemical injection systems can be used in most types of structure, although flint walls and rubble infilled walls can be difficult to treat.

Figure 6.24 A plumbing leak has caused deterioration of the chipboard deck laid on expanded polystyrene. An unsealed joint in the polyethylene VCL allowed moisture to rise to the surface

Excluding rising damp

Rising damp can occur in older buildings, many of which don't have DPCs in external walls. Fortunately, many of them have suspended floors, only the perimeters of which are at risk. BS CP 102 was for many years the authoritative source of information on construction standards relating to rising damp, and has even now only been partially replaced by BS 8102 and BS 8215.

DPMs

The DPM in a groundbearing floor slab can be laid in a variety of positions. With a fully bonded screed, the DPM must be laid under the slab; for an existing screed, a surface DPM can be used. The exception is an epoxy DPM, which can act as a bonding agent and can be laid as a sandwich.

Existing dwellings being rehabilitated will often have solid floors which do not have DPMs. Many houses built between 1950 and 1966 had floors which were finished with thermoplastic tiles stuck down with a bitumen adhesive. This system tolerated moderately damp conditions, so it was common not to provide a DPM in the base. If the flooring is removed, make an assessment of the moisture condition of the base as a DPM may be required before laying moisture sensitive flooring.

Moisture-sensitive materials, such as chipboard, other timber products, flexible PVC, linoleum, or cork tiles, should be laid only on a floor which has a satisfactory DPM. In a survey of rehabilitation in progress, faults included failure to provide a DPM, even in vulnerable situations, and unsatisfactory linking of DPMs with DPCs.

Where timber ground floors have been replaced by solid floors, BRE has been commissioned frequently to investigate rising damp showing on the external, usually solid, wall – Figure 6.25.

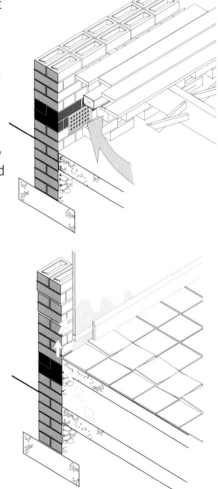

Figure 6.25 There is a risk of rising damp where a solid floor replaces a suspended floor, unless an adequate DPM is installed
(top) before replacement ...
(bottom) ... and after replacement

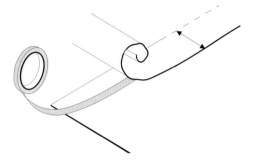

Figure 6.26 Sticking the overlap of a DPM

DPMs must be continuous, above or below the base and linked to the DPC in the walls. Polyethylene below the base should be at least 300μm or 250μm if the product has a BBA certificate or is to the PIFA standard. Where sheet material DPMs are being laid below replacement floor slabs, it is important to ensure adequacy in the joints between sheets. The preferred method of forming the joint is to overlap the sheets by at least 150 mm, and stick the joint with double-sided pressure-sensitive tape – Figure 6.26.

If welts are used to join sheets, construct them in a four-stage operation – Figure 6.27. Hold the welt flat until the slab or screed is placed.

Service entry points can be a source of penetration of damp – Figure 6.28, mainly in buildings built in ground with high water tables, or where persistent dampness is present in the hardcore. When replacing slabs in such locations, it is worth taking care with damp-proofing service entry points – Figure 6.29.

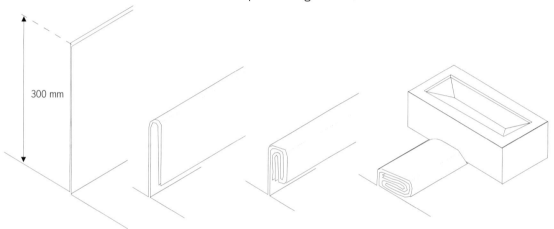

300 mm

Figure6.27 Forming a welt in a DPM

first fold *second fold* *third fold weighted until screed is laid*

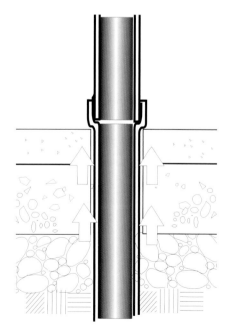

Figure 6.28 Service installations are the most likely points at which rising damp will show

Brush-applied and hot-poured materials are available for membranes laid on top of the base. They should be applied to provide a dry film, at least 0.6mm thick. Take care selecting these materials as some of the solvents may not be compatible with foamed plastics insulants. In any case, allow sufficient time for any solvent to evaporate before covering. Consider the possibility of poor workmanship with brush-applied materials. They must be covered with a sand/cement screed not less than 50 mm thick.

Another alternative is a waterproof flooring, such as mastic asphalt to BS 6925, which will add around 20 mm to the finished floor level; it is laid in accordance with BS 8204: Part 5.

The use of sandwich membranes is restricted to where the whole floor is being replaced; in this case it will be beneficial, given adequate headroom, to install thermal insulation below the slab.

Surface DPMs based on proprietary epoxy resin systems have been available since around 1965. They have a good track record, but have only recently become widely accepted because of their high cost. They are rarely considered

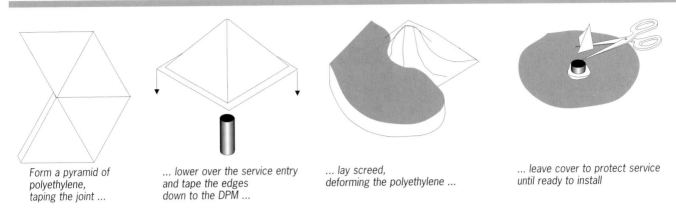

*Form a pyramid of
polyethylene,
taping the joint ...*

*... lower over the service entry
and tape the edges
down to the DPM ...*

*... lay screed,
deforming the polyethylene ...*

*... leave cover to protect service
until ready to install*

Figure 6.29 One method of protecting a service entry point

Figure 6.30 The base being prepared for casting a raft for a multi-storey building. Column bases are already in position

Figure 6.31 Laying a DPM which also acts as blinding to prevent fines migrating to the hardcore. Taping provides integrity to the seal

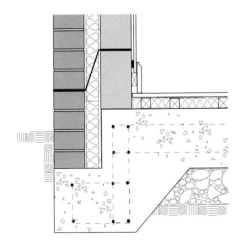

Figure 6.32 Insulation should be started as low as possible at the toe of concrete rafts

for new work but are most useful for renovation and change of use where no DPM is present. They can be applied to clean surfaces of concrete or screed, and should be covered by a latex levelling compound at least 3 mm thick. These surface DPMs are also useful for controlling excess constructional water in thick constructions when it would be uneconomic to wait for them to dry out.

Protection of a raft from rising damp has to be done underneath the concrete, in order to provide continuity. This means that some of the geometry is complicated, particularly at perimeters and thickenings – Figures 6.30 and 6.31

There is a risk of condensation occurring at thermal bridges with this kind of floor. The thermal insulation in the cavity of an external wall cannot be taken sufficiently low to overlap that within the floor – Figure 6.32.

Magnesite flooring

Magnesite can only be used in dry situations. These floorings are very vulnerable to dampness, and those without adequate DPMs will probably have failed and have been removed long ago if covered by an impervious flooring. If it remains damp, the oxychloride reaction is reversed; the material rapidly loses strength – Figure 6.33 – and in the worst cases may form a mush. The reversed reaction releases magnesium chloride which rapidly corrodes metals and can enter concrete bases and affect reinforcement.

Water can reach magnesite in a number of ways: spillage, plumbing leaks, and construction water during major refurbishment of buildings. However, it was not uncommon for it to be laid in ground floors without any DPM. Any rising moisture can diffuse through the flooring and evaporate away without harm. However, if such a floor is covered by another impervious flooring, such as PVC or rubber-backed carpet, moisture can build up in the magnesite. Bear in mind that such screeds should not be covered with a new DPM either, since in these circumstances they will suffer accelerated deterioration. All such screeds should be completely removed before installing new surface DPMs.

Magnesite can 'sweat' in moderately humid conditions. This is characterised by beads of magnesium chloride forming on the surface, and is not the result of condensation. Sweating is to some extent a property of the material since magnesium chloride, an essential ingredient, is very hygroscopic. Because it is in slight excess, it readily takes up moisture from humid air. Magnesium chloride may migrate into adjacent walls and will cause dampness there because of its hygroscopicity.

The slight excess of magnesium chloride also ensures that magnesite flooring is electrically conducting. As a result, moisture meters of the resistance type cannot be used to assess the moisture condition of this flooring. Even when bone dry, most meters give nearly a full-scale deflection – Figure 6.34.

Figure 6.33 Magnesite flooring breaking up following prolonged wetting

Figure 6.34 Full scale deflection of an electrical resistance moisture meter on a bone dry sample from a two coat magnesite floor. Wet or dry, the meter would give the same reading

long to dry. Drying is best effected from the outer surface, though rendered finishes take longer.

Stone walls are likely to dry out more rapidly than those of brick, since they are often less porous. However, rubble cores in the walls may have to be drained just like cavities.

These points are important:
- ❑ keep warm air flowing through the building by heating and ventilation;
- ❑ keep windows and doors kept open to give good ventilation, even when the heating is on; take precautions against housebreaking;
- ❑ remove loose floor coverings and carpets for drying and storage;
- ❑ lift floorboards, especially near walls, to increase draught under suspended floors; this includes upper floors affected by the flooding;
- ❑ strip impervious wallcoverings to help the walls dry out;
- ❑ keep furniture and pictures away from damp walls;
- ❑ keep cupboard doors open.

If the heating system is still operable, set the thermostat to 22° or above, with as much ventilation as practicable.

Inspections

Disconnect and test all electrical installations that have been immersed in water. At the same time, examine and test all appliances: this includes boilers, heaters and cookers. Cables in good condition should not be affected by immersion, but junctions certainly will be.

Open ducts or conduits containing cables to help drain any trapped water. Once the cleaning and drying is completed, arrange for the installation to be tested for earth continuity and insulation resistance as laid down in current IEE Regulations, and issue an inspection certificate. Inspect the electricity installation every month for the first six months after the initial test, and at least twice again in the following six months.

Remedies

Walls

During cold weather, wet walls may be damaged by frost, causing the surface of the brickwork to crack and powder away. Some walls may expand because of the dampness and then contract on drying, producing fine cracks which usually can be dealt with simply by repainting. Some types of wall plaster soften readily when wet and crumble when they dry out again. Others may expand or contract such that replacement is needed.

Timber

Probably the greatest danger caused by flood waters, even months after inundation, is rot in timber. The longer timbers remain wet the more likely they are to rot. It is worth keeping floorboards raised and even introducing special dehumidifying equipment to dry out floors as quickly as possible. Splitting can be minimised by ensuring the timbers dry on both faces, which is helped by removing panelling and skirtings.

All timbers, including door frames, the ends of joists, skirting boards and floorboards, attached to or embedded in damp walls are vulnerable and should be moved away or cut back from the walls. Where joists are embedded in brickwork, supporting metal hangers may be needed instead.

Salt contamination

Salt contamination from sea water or other sources affects the readings from an electrical moisture meter. If it is possible to accurately measure moisture content of timber, values should be 24% or below when measured from October to May and below 22% for the remainder of the year When the timbers have dried, inspect the underfloor timbers six months later and then again in a year's time. Several types of rot can affect the timbers, which show up as brown or white strands, small orange or white blotches, splitting of the timber and softness nearby and, in extreme cases, the growth of fungus – see *Recognising wood rot and insect damage in buildings* and *Remedial treatment of wood rot and insect attack in buildings.*

Floors

When tongued and grooved floorboards and sheet floor coverings have dried, they may shrink leaving gaps between the boards which may need to be filled or tightened up. Some types of chipboard used for flooring will have swollen during wetting and become permanently weak as a result. The only option here is to replace the material, preferably with one that remains stable when damp.

Wood blocks and other coverings, such as vinyl or linoleum, which have been stuck to floorboards or concrete screeds may have lifted because the adhesive has weakened on wetting. Some materials, particularly wood block and strip, may have swollen and become damaged.

Impervious floor coverings should not be laid until drying out is acceptably complete. Test using a hygrometer: readings should be in the range 75 – 80%. Valuable timber panelling must be dried thoroughly and not replaced until the backing walls are completely dry. In timber framed walls, take away some plasterboard panels to expose the timber framework and remove the insulation removed as necessary; don't replace the wall linings until drying is complete.

Doors

Panelled doors are unlikely to be affected seriously unless the panels are made of the type of plywood which expands because the adhesive is sensitive to water. Modern flush doors are often more severely damaged and require replacement. Other doors and windows may stick but should not be eased by planing the edges until drying and shrinkage is complete.

Metals

Metals are likely to have escaped serious damage unless seawater flooding has occurred. Steel reinforcement embedded in concrete may corrode and expand causing long-term damage. In other cases, the drying process should leave the metals unscathed, although locks and hinges should be oiled to prevent them rusting and seizing up.

Redecoration

Delay redecoration until the walls have dried thoroughly and use porous coatings, such as emulsion paint, rather than wallpaper. Treat walls with a fungicide if there are signs of mould growth.

Figure 7.4 Persistent leakage from a warning pipe. The warning has gone unheeded

Complete list of contents

REFERENCES AND FURTHER READING

Building Regulations England and Wales. Approved Document C4. Site preparation and resistance to moisture. London, The Stationery Office.

Building Regulations England and Wales. Approved Document L. Conservation of fuel and power. London, The Stationery Office.

Technical Standards for compliance with the Building Standards (Scotland) Regulations 1990: Part G: Preparation of sites, resistance to moisture and resistance to condensation. London, The Stationery Office.

Clean Air Act 1956. London, The Stationery Office.

Public Health Act 1875. London, The Stationery Office.

Public Health Act 1936. London, The Stationery Office.

Robust details: Limiting thermal bridging and air leakage - robust construction details for dwellings and similar buildings. London, The Stationery Office.

English house condition survey 1996. Department of the Environment. London, The Stationery Office, 1998.

Scottish house condition survey 1996. Scottish Homes. Survey report. Edinburgh, Scottish Homes, 1997.

Working with asbestos cement. HSE Guidance Note EH36.

MATCH software.
http://www.match-box.dk/dk/hvaderbeskrivelse.htm

Energy Efficiency Best Practice programme (EEBPp)
Good Practice Guides. Available from:
0800 585749   energy-efficiency.gov.uk
174 Minimising thermal bridging in new dwellings
183 Minimising thermal bridging when upgrading existing housing

Rain Penetration Investigations - A summary of the findings of CIB Working Commission on Rain Penetration - Oslo 1963.

The hole truth. Cook A. Published in *Building*.

EC Concerted Action of Indoor Air Quality and its Impact on Man

Directives for the Assessment of Manufactured Plastics Floorings

Dampness in cob walls. Trotman, PM. Conference paper: Out of earth II at Centre for Earthen Architecture, Plymouth School of Architecture, May 1995.

Injection systems for damp-proofing. Sharpe, RW. Building and Environment Vol 12, 191-197. Pergamon 1977.

British Wood Preserving and Damp-Proofing Association (BWPDA)
www.bwpda.co.uk

British Board of Agrément (BBA)
www.bbacerts.co.uk

Housing Association Property Mutual (HAPM)
www.hapm.co.uk

BRE publications

Rain penetration through masonry walls: diagnosis and remedial measures BR117

Assessing traditional housing for rehabilitation BR167

Understanding and improving the weathertightness of large panel system dwellings BR 214

Recognising wood rot and insect damage in buildings BR232

Housing design handbook BR253

Remedial treatment of wood rot and insect damage in buildings BR 256

Thermal insulation: avoiding risks BR262

Cracking in buildings BR292

Roofs and roofing BR302

Walls, windows and doors BR352

Building services BR404

Foundations, basements and external works BR440

Floors and flooring BR460

Potential implications of climate change in the built environment FBE2

Principles of modern building. London, 1938, The Stationery Office

Current Papers
5/83 Trussed rafter roofs
81/74 Some observations on the behaviour of weather protective features on external walls
90/74 New ways with waterproof joints

Information Papers
16/03 Proprietry renders
3/93 Quality in new-build housing

Defect Action Sheets
9 Pitched roofs: sarking felt underlay – drainage from roof
33 Flat roofs: built-up bitumen felt – remedying rain penetration
114 Slated and tiled pitched roofs; flashings and cavity trays for step and stagger layouts – specification

Good Building Guides
3 Damp-proofing existing basements
18 Choosing external rendering
47 Level external thresholds: reducing moisture penetration and thermal bridging

Good Repair Guides
11 Repairing flood damage (in four parts)
23 Treating dampness in basements
33 Assessing moisture in building materials *(in 3 parts)*

Digests
54 Damp-proofing solid floors
89 Sulphate attack on brickwork
163 Drying out buildings
329 Installing wall ties in existing construction
364 Design of timber floors to prevent decay
370 Control of lichens, moulds and similar growths
377 Selecting windows by performance
380 Damp-proof courses
401 Replacing wall ties
440 Weathering of white external PVC-U

British Standards Institution

BS 743: 1970 Specification for materials for damp-proof courses

BS 747: 2000 Reinforced bitumen sheets for roofing

BS 1097 Specification for mastic asphalt for building (limestone aggregate)

BS 1178: 1982 Specification for milled lead sheet for building purposes

BS 1199: 1976 Specifications for building sands from natural sources

BS 1418: 1973 Specification for mastic asphalt for building (natural rock asphalt aggregate)

BS 2870: 1980 Specification for rolled copper and copper alloys: sheet, strip and foil

BS 3921: 1985 Specification for clay bricks

BS 4016: 1997 Specification for flexible building membranes (breather type

BS 5250: 2002 Code of practice for control of condensation in buildings

BS 5493: 1977 Code of practice for protective coating of iron and steel structures against corrosion

BS 5628-3: 2001 Code of practice for use of masonry. Materials and components, design and workmanship

BS 5368 Methods of testing windows

BS 5617: 1985 Specification for urea-formaldehyde (UF) foam systems suitable for thermal insulation of cavity walls with masonry or concrete inner and outer leaves

BS 5618: 1985 Code of practice for thermal insulation of cavity walls (with masonry or concrete inner and outer leaves) by filling with urea-formaldehyde (UF) foam systems

BS 5628-3: 2001 Code of practice for use of masonry. Materials and components, design and workmanship

BS 5889: 1989 Specification for one-part gun grade silicone-based sealant

BS 6232-1: 1982 Thermal insulation of cavity walls by filling with blown man-made mineral fibre. Specification for the performance of installation systems

BS 6232-2: 1982 Thermal insulation of cavity walls by filling with blown man-made mineral fibre. Code of practice for installation of blown man-made mineral fibre in cavity walls with masonry and/or concrete leaves

BS 6367: 1983 Code of practice for drainage of roofs and paved areas

BS 6375 Performance of windows

BS 6398: 1983 Specification for bitumen damp-proof courses for masonry

BS 6477: 1992 Specification for water repellents for masonry surfaces

BS 6515: 1984 Specification for polyethylene damp-proof courses for masonry

BS 6576: 1985 Code of practice for installation of chemical damp-proof courses

BS 6577: 1985 Specification for mastic asphalt for building (natural rock asphalt aggregate)

BS 6925: 1988 Specification for mastic asphalt for building and civil engineering (limestone aggregate)

BS 8102: 1990 Code of practice for protection of structures against water from the ground

BS 8104: 1992 Code of practice for assessing exposure of walls to wind-driven rain

BS 8200: 1985 Code of practice for design of non-loadbearing external vertical enclosures of buildings

BS 8203: 2001 Code of practice for installation of resilient floor coverings

BS 8204-5: 1994 Screeds, bases and in-situ floorings. Code of practice for mastic asphalt underlays and wearing surfaces

BS 8215: 1991 Code of practice for design and installation of damp-proof courses in masonry construction

BS EN 12371: 2001 Natural stone test methods. Determination of frost resistance

BS EN 13755: 2002 Natural stone. Test methods. Determination of water absorption at atmospheric pressure

BS EN ISO 12572:2002 Hygrothermal performance of building materials – Determination of moisture content by drying at elevated temperature

BS EN ISO 13788:2002 Hygrothermal performance of building components and building elements – Internal surface temperature to avoid critical surface humidity and interstitial condensation – Calculation methodsBS CP 102: 1973 Code of practice for protection of buildings against water from the ground

BS CP 144: Part 4: 1970 Roof coverings. Mastic asphalt